Foundations of Risk Analysis

Foundations of Risk Analysis

A Knowledge and Decision-Oriented Perspective

Terje Aven
University of Stavanger, Norway

John Wiley & Sons, Ltd

Copyright © 2003   John Wiley & Sons Ltd, The Atrium, Southern Gate, Chichester,
West Sussex PO19 8SQ, England

Telephone (+44) 1243 779777

Email (for orders and customer service enquiries): cs-books@wiley.co.uk
Visit our Home Page on www.wileyeurope.com or www.wiley.com

All Rights Reserved. No part of this publication may be reproduced, stored in a retrieval system or transmitted in any form or by any means, electronic, mechanical, photocopying, recording, scanning or otherwise, except under the terms of the Copyright, Designs and Patents Act 1988 or under the terms of a licence issued by the Copyright Licensing Agency Ltd, 90 Tottenham Court Road, London W1P 4LP, UK, without the permission in writing of the Publisher. Requests to the Publisher should be addressed to the Permissions Department, John Wiley & Sons Ltd, The Atrium, Southern Gate, Chichester, West Sussex PO19 8SQ, England, or emailed to permreq@wiley.co.uk, or faxed to
(+44) 1243 770620.

This publication is designed to provide accurate and authoritative information in regard to the subject matter covered. It is sold on the understanding that the Publisher is not engaged in rendering professional services. If professional advice or other expert assistance is required, the services of a competent professional should be sought.

Other Wiley Editorial Offices

John Wiley & Sons Inc., 111 River Street, Hoboken, NJ 07030, USA

Jossey-Bass, 989 Market Street, San Francisco, CA 94103-1741, USA

Wiley-VCH Verlag GmbH, Boschstr. 12, D-69469 Weinheim, Germany

John Wiley & Sons Australia Ltd, 33 Park Road, Milton, Queensland 4064, Australia

John Wiley & Sons (Asia) Pte Ltd, 2 Clementi Loop #02-01, Jin Xing Distripark, Singapore 129809

John Wiley & Sons Canada Ltd, 22 Worcester Road, Etobicoke, Ontario, Canada M9W 1L1

Wiley also publishes its books in a variety of electronic formats. Some content that appears in print may not be available in electronic books.

British Library Cataloguing in Publication Data

A catalogue record for this book is available from the British Library

ISBN 0-471-49548-4

Typeset in 10/12pt Times by Laserwords Private Limited, Chennai, India
Printed and bound in Great Britain by Antony Rowe Ltd, Chippenham, Wiltshire
This book is printed on acid-free paper responsibly manufactured from sustainable forestry in which at least two trees are planted for each one used for paper production.

Contents

Preface   **ix**

1 Introduction   **1**
   1.1 The Importance of Risk and Uncertainty Assessments   1
   1.2 The Need to Develop a Proper Risk Analysis Framework   4
   Bibliographic Notes   6

2 Common Thinking about Risk and Risk Analysis   **7**
   2.1 Accident Risk   7
      2.1.1 Accident Statistics   7
      2.1.2 Risk Analysis   11
      2.1.3 Reliability Analysis   24
   2.2 Economic Risk   28
      2.2.1 General Definitions of Economic Risk in Business and Project Management   28
      2.2.2 A Cost Risk Analysis   30
      2.2.3 Finance and Portfolio Theory   31
      2.2.4 Treatment of Risk in Project Discounted Cash Flow Analysis   34
   2.3 Discussion and Conclusions   36
      2.3.1 The Classical Approach   36
      2.3.2 The Bayesian Paradigm   37
      2.3.3 Economic Risk and Rational Decision-Making   39
      2.3.4 Other Perspectives and Applications   40
      2.3.5 Conclusions   42
   Bibliographic Notes   43

3 How to Think about Risk and Risk Analysis   **47**
   3.1 Basic Ideas and Principles   47
      3.1.1 Background Information   50
      3.1.2 Models and Simplifications in Probability Considerations   51
      3.1.3 Observable Quantities   51
   3.2 Economic Risk   52
      3.2.1 A Simple Cost Risk Example   52
      3.2.2 Production Risk   55

	3.2.3	Business and Project Management	57
	3.2.4	Investing Money in a Stock Market	58
	3.2.5	Discounted Cash Flow Analysis	59
3.3	Accident Risk		60
Bibliographic Notes			62

4 How to Assess Uncertainties and Specify Probabilities 63

4.1	What Is a Good Probability Assignment?		64
	4.1.1	Criteria for Evaluating Probabilities	64
	4.1.2	Heuristics and Biases	66
	4.1.3	Evaluation of the Assessors	67
	4.1.4	Standardization and Consensus	68
4.2	Modelling		68
	4.2.1	Examples of Models	69
	4.2.2	Discussion	70
4.3	Assessing Uncertainty of Y		71
	4.3.1	Assignments Based on Classical Statistical Methods	72
	4.3.2	Analyst Judgements Using All Sources of Information	73
	4.3.3	Formal Expert Elicitation	74
	4.3.4	Bayesian Analysis	75
4.4	Uncertainty Assessments of a Vector $\mathbf{X}$		83
	4.4.1	Cost Risk	83
	4.4.2	Production Risk	85
	4.4.3	Reliability Analysis	86
4.5	Discussion and Conclusions		90
Bibliographic Notes			92

5 How to Use Risk Analysis to Support Decision-Making 95

5.1	What Is a Good Decision?		96
	5.1.1	Features of a Decision-Making Model	97
	5.1.2	Decision-Support Tools	98
	5.1.3	Discussion	103
5.2	Some Examples		106
	5.2.1	Accident Risk	106
	5.2.2	Scrap in Place or Complete Removal of Plant	108
	5.2.3	Production System	113
	5.2.4	Reliability Target	114
	5.2.5	Health Risk	116
	5.2.6	Warranties	119
	5.2.7	Offshore Development Project	120
	5.2.8	Risk Assessment: National Sector	122
	5.2.9	Multi-Attribute Utility Example	124
5.3	Risk Problem Classification Schemes		127
	5.3.1	A Scheme Based on Potential Consequences and Uncertainties	127

	5.3.2	A Scheme Based on Closeness to Hazard and Level of Authority	131

Bibliographic Notes ... 142

6 Summary and Conclusions — 145

Appendix A Basic Theory of Probability and Statistics — 149
A.1 Probability Theory ... 149
    A.1.1 Types of Probabilities ... 149
    A.1.2 Probability Rules ... 151
    A.1.3 Random Quantities (Random Variables) ... 155
    A.1.4 Some Common Discrete Probability Distributions (Models) ... 159
    A.1.5 Some Common Continuous Distributions (Models) ... 160
    A.1.6 Some Remarks on Probability Models and Their Parameters ... 164
    A.1.7 Random Processes ... 165
A.2 Classical Statistical Inference ... 166
    A.2.1 Non-Parametric Estimation ... 166
    A.2.2 Estimation of Distribution Parameters ... 167
    A.2.3 Testing Hypotheses ... 169
    A.2.4 Regression ... 170
A.3 Bayesian Inference ... 171
    A.3.1 Statistical (Bayesian) Decision Analysis ... 173
Bibliographic Notes ... 174

Appendix B Terminology — 175

Bibliography — 179

Index — 187

Preface

This book is about foundational issues in risk and risk analysis; how risk should be expressed; what the meaning of risk is; how to understand and use models; how to understand and address uncertainty; and how parametric probability models like the Poisson model should be understood and used. A unifying and holistic approach to risk and uncertainty is presented, for different applications and disciplines. Industry and business applications are highlighted, but aspects related to other areas are included. Decision situations covered include concept optimization and the need for measures to reduce risk for a production system, the choice between alternative investment projects and the use of a type of medical treatment.

My aim is to give recommendations and discuss how to approach risk and uncertainty to support decision-making. We go one step back compared to what is common in risk analysis books and papers, and ask how we should think at an early phase of conceptualization and modelling. When the concepts and models have been established, we can use the well-defined models covered thoroughly by others.

Here are the key principles of the recommended approach. The focus is on so-called observable quantities, that is, quantities expressing states of the 'world' or nature that are unknown at the time of the analysis but will (or could) become known in the future; these quantities are predicted in the risk analysis and probability is used as a measure of uncertainty related to the true values of these quantities. Examples of observable quantities are production volume, production loss, the number of fatalities and the occurrence of an accident.

These are the main elements of the unifying approach. The emphasis on these principles gives a framework that is easy to understand and use in a decision-making context. But to see that these simple principles are in fact the important ones, has been a long process for me. It started more than ten years ago when I worked in an oil company where I carried out a lot of risk and reliability analyses to support decision-making related to choice of platform concepts and arrangements. I presented risk analysis results to management but, I must admit, I had no proper probabilistic basis for the analyses. So when I was asked to explain how to understand the probability and frequency estimates, I had problems. Uncertainty in the estimates was a topic we did not like to speak about as we could not deal with it properly. We could not assess or quantify the uncertainty, although we had to admit that it was considerably large in most cases; a factor of 10 was often indicated, meaning that the true risk could be either a factor 10 above or below the estimated value. I found this discussion of uncertainty frustrating and disturbing. Risk analysis should be a tool for dealing with uncertainty, but by the way we were thinking, I felt that the analysis in a way created uncertainty that was not inherent in the system being analysed. And that could not be right.

As a reliability and risk analyst, I also noted that the way we were dealing with risk in this type of risk analysis was totally different from the one adopted when predicting the future gas and oil volumes from production systems. Then focus was not on estimating some true probability and risk numbers, but predicting observable quantities such as production volumes and the number of failures. Uncertainty was related to the ability to predict a correct value and it was expressed by probability distributions of the observable quantities, which is in fact in lines with the main principles of the recommended approach of this book.

I began trying to clarify in my own mind what the basis of risk analysis should be. I looked for alternative ways of thinking, in particular the Bayesian approach. But it was not easy to see from these how risk and uncertainty should be dealt with. I found the presentation of the Bayesian approach very technical and theoretical. A subjective probability linked to betting and utilities was something I could not use as a cornerstone of my framework. Probability and risk should be associated with uncertainty, not our attitude to winning or losing money as in a utility-based definition. I studied the literature and established practice on economic risk, project management and finance, and Bayesian decision analysis, and I was inspired by the use of subjective probabilities expressing uncertainty, but I was somewhat disappointed when I looked closer into the theories. References were made to some literature restricting the risk concept to situations where the probabilities related to future outcomes are known, and uncertainty for the more common situations of unknown probabilities. I don't think anyone uses this convention and I certainly hope not. It violates the intuitive interpretation of risk, which is closely related to situations of unpredictability and uncertainty. The economic risk theory appreciates subjectivity but in practice it is difficult to discern the underlying philosophy. Classical statistical principles and methods are used, as well as Bayesian principles and methods. Even more frustrating was the strong link between uncertainty assessments, utilities and decision-making. To me it is essential to distinguish between what I consider to be decision support, for example the results from risk analyses, and the decision-making itself.

The process I went through clearly demonstrated the need to rethink the basis of risk analysis. I could not find a proper framework to work in. Such a framework should be established. The framework should have a clear focus and an understanding of what can be considered as technicalities. Some features of the approach were evident to me. Attention should be placed on observable quantities and the use of probability as a subjective measure of uncertainty. First comes the world, the reality (observable quantities), then uncertainties and finally probabilities. Much of the existing classical thinking on risk analysis puts probabilities first, and in my opinion this gives the wrong focus. The approach to be developed should make risk analysis a tool for dealing with uncertainties, not create uncertainties and in that way disturb the message of the analysis. This was the start of a very interesting and challenging task, writing this book.

The main aim of this book is to give risk analysts and others an authoritative guide, with discussion, on how to approach risk and uncertainty when the basis is subjective probabilities, expressing uncertainty, and the rules of probability. How should a risk analyst think when he or she is planning and conducting a risk analysis? And here are some more specific questions:

- How do we express risk and uncertainty?
- How do we understand a subjective probability?
- How do we understand and use models?
- How do we understand and use parametric distribution classes and parameters?
- How do we use historical data and expert opinions?

Chapters 3 to 6 present an approach or a framework that provides answers to these questions, an approach that is based on some simple ideas or principles:

- Focus is placed on quantities expressing states of the 'world', i.e. quantities of the physical reality or nature that are unknown at the time of the analysis but will, if the system being analysed is actually implemented, take some value in the future, and possibly become known. We refer to these quantities as *observable* quantities.
- The observable quantities are predicted.
- Uncertainty related to what values the observable quantities will take is expressed by means of probabilities. This uncertainty is *epistemic*, i.e. a result of lack of knowledge.
- Models in a risk analysis context are deterministic functions linking observable quantities on different levels of detail. The models are simplified representations of the world.

The notion of an observable quantity is to be interpreted as a potentially observable quantity; for example, we may not actually observe the number of injuries (suitably defined) in a process plant although it is clearly expressing a state of the world. The point is that a true number exists and if sufficient resources were made available, that number could be found.

Placing attention on the above principles would give a unified structure to risk analysis that is simple and in our view provides a good basis for decision-making. Chapter 3 presents the principles and gives some examples of applications from business and engineering. Chapter 4 is more technical and discusses in more detail how to use probability to express uncertainty. What is a good probability assignment? How do we use information when assigning our probabilities? How should we use models? What is a good model? Is it meaningful to talk about model uncertainty? How should we update our probabilities when new information becomes available? And how should we assess uncertainties of 'similar units', for example pumps of the same type? A full Bayesian analysis could be used, but in many cases a simplified approach for assessing the uncertainties is needed, so that we can make the probability assignments without adopting the somewhat sophisticated procedure of specifying prior distributions of parameters. An example is the initiating event and the branch events in an event tree where often direct probability assignments are preferred instead of using the full Bayesian procedure with specification of priors of the branch probabilities and the occurrence rate of the initiating event. Guidance is given on when to use such a simple approach and when to run a complete Bayesian analysis. It has been essential for us to provide a simple assignment process that works in practice for the number of probabilities and probability distributions in a risk analysis. We should not introduce distribution classes with unknown parameters when not required. Furthermore, meaningful interpretations must be given to the distribution classes and the parameters whenever they are used. There is no point in speaking about uncertainty of parameters unless they are observable, i.e. not fictional.

The literature in mathematics and philosophy discusses several approaches for expressing uncertainty. Examples are possibility theory and fuzzy logic. This book does not discuss the various approaches; it simply states that probability and probability calculus are used as the sole means for expressing uncertainty. We strongly believe that probability is the most suitable tool. The interpretation of probability is subject to debate, but its calculus is largely universal.

Chapter 5 discusses how to use risk analysis to support decision-making. What is a good decision? What information is required in different situations to support decision-making? Examples of decision-making challenges are discussed. Cost-benefit analyses and Bayesian decision analyses can be useful tools in decision-making, but in general we recommend a flexible approach to decision-making, in which uncertainty and uncertainty assessments (risk) provide decision support but there is no attempt to explicitly weight future outcomes or different categories of risks related to safety, environmental issues and costs. The main points of Chapters 3 to 5 are summarized in Chapter 6.

Reference is above given to the use of subjective probability. In applications the word 'subjective', or related terms such as 'personalistic', is often difficult as it seems to indicate that the results you present as an analyst are subjective whereas adopting an alternative risk analysis approach can present objective results. So why should we always focus on the subjective aspects when using our approach? In fact, all risk analysis approaches produce subjective risk results; the only reason for using the word 'subjective' is that this is its original, historical name. We prefer to use 'probability as a measure of uncertainty' and make it clear who is the assessor of the uncertainty, since this is the way we interpret a subjective probability and we avoid the word 'subjective'.

In our view, teaching the risk analyst how to approach risk and uncertainty cannot be done without giving a context for the recommended thinking and methods. What are the alternative views in dealing with risk and uncertainty?

This book aims to review and discuss common thinking about risk and uncertainty, and relate it to the presentation of Chapters 3 to 6. Chapter 2, which covers this review and discussion, is therefore important in itself and an essential basis for the later chapters. It comes after Chapter 1, which discusses the need for addressing risk and uncertainty and the need for developing a proper risk analysis framework.

The book covers four main directions of thought:

- The classical approach with focus on best estimates. Risk is considered a property of the system being analysed and the risk analysis provides estimates of this risk.
- The classical approach with uncertainty analysis, also known as the probability of frequency framework. Subjective probability distributions are used to express uncertainty of the underlying true risk numbers.
- The Bayesian approach as presented in the literature.
- Our predictive approach, which may be called a predictive Bayesian approach.

Chapter 2 presents the first two approaches (Sections 2.1 and 2.2), and relates them to Bayesian thinking (Section 2.3), whereas Chapters 3 to 6 present our predictive approach. The presentation in Chapters 4 and 5 also cover key aspects of the Bayesian paradigm (Chapter 4) and Bayesian decision theory (Chapter 5), as these are basic elements of our predictive approach. To obtain a complete picture of how these different perspectives are related, Chapters 2 to 6 need to be read carefully.

This book is written primarily for risk analysts and other specialists dealing with risk and risk analysis, as well as academics and graduates. Conceptually it is rather challenging. To quickly appreciate the book, the reader should be familiar with basic probability theory. The key statistical concepts are introduced and discussed thoroughly in the book, as well as some basic risk analysis tools such as fault trees and event trees. Appendix A summarizes some basic probability theory and statistical analysis. This makes the book more self-contained, gives it the required sharpness with respect to relevant concepts and tools, and makes it accessible to readers outside the primary target group. The book is based on and relates to the research literature in the field of risk and uncertainty. References are kept to a minimum throughout, but bibliographic notes at the end of each chapter give a brief review of the material plus relevant references.

Most of the applications in the book are from industry and business, but there are some examples from medicine and criminal law. However, the ideas, principles and methods are general and applicable to other areas. What is required is an interest in studying phenomena that are uncertain at the time of decision-making, and that covers quite a lot of disciplines.

This book is primarily about how to approach risk and uncertainty, and it provides clear recommendations and guidance. But it is not a recipe book telling you how to plan, conduct and use risk analysis in different situations. For example, how should a risk analysis of a large process plant be carried out? How should we analyse the development of a fire scenario? How should we analyse the evacuation from the plant? These issues are not covered. What it does cover are the general thinking process related to risk and uncertainty quantification, and the probabilistic tools to achieve it. When referring to our approach as a unifying framework, this relates only to these overall features. Within each discipline and area of application there are several tailor-made risk analysis methods and procedures.

The terminology used in this book is summarized in Appendix B. It is largely in line with the ISO standard on risk management terminology (ISO 2002).

We believe this book is important as it provides a guide on how to approach risk and uncertainty in a practical decision-making context and it is precise on concepts and tools. The principles and methods presented should work in practice. Consequently, we have put less emphasis on Bayesian updating procedures and formal decision analysis than perhaps would have been expected when presenting an approach to risk and uncertainty based on the use of subjective probabilities. Technicalities are reduced to a minimum, ideas and principles are highlighted.

Our approach means a humble attitude to risk and the possession of the truth, and hopefully it will be more attractive to social scientists and others, who have strongly criticized the prevailing thinking of risk analysis and evaluation in the engineering environment. We agree that a sharp distinction between objective, real risk and perceived risk cannot be made. Risk is primarily a judgement, not a fact. To a large extent, our way of thinking integrates technical and economic risk analyses and social science perspectives on risk. As risk expresses uncertainty about the world, risk perception has a role to play in guiding decision-makers. Professional risk analysts do not have the exclusive right to describe risk.

Scientifically, our perspective on uncertainty and risk can be classified as instrumental, in the sense that we see the risk analysis methods and models as nothing more than useful instruments for getting insights about the world and to support decision-making. Methods and models are not appropriately interpreted as being true or false.

Acknowledgements Several people have provided helpful comments on portions of the manuscript at various stages. In particular, I would like to acknowledge Sigve Apeland, Gerhard Ersdal, Uwe Jensen, Vidar Kristensen, Henrik Kortner, Jens Kørte, Espen Fyhn Nilsen, Ove Njå, Petter Osmundsen, Kjell Sandve and Jan Erik Vinnem. I especially thank Tim Bedford, University of Strathclyde, and Bent Natvig, University of Oslo, for the great deal of time and effort they spent reading and preparing comments. Over the years, I have benefited from many discussions with a number of people, including Bo Bergman, Roger Cooke, Jørund Gåsemyr, Nozer Singpurwalla, Odd Tveit, Jørn Vatn and Rune Winther. I would like to make special acknowledgment to Dennis Lindley and William Q. Meeker for their interest in my ideas and this book; their feedback has substantially improved parts of it. Thanks also go to the many formal reviewers for providing advice on content and organization. Their informed criticism motivated several refinements and improvements. I take full responsibility for any errors that remain.

For financial support, I thank the University of Stavanger, the University of Oslo and the Norwegian Research Council.

I also acknowledge the editing and production staff at John Wiley & Sons for their careful work. In particular, I appreciate the smooth cooperation of Sharon Clutton, Rob Calver and Lucy Bryan.

1

Introduction

1.1 THE IMPORTANCE OF RISK AND UNCERTAINTY ASSESSMENTS

The concept of risk and risk assessments has a long history. More than 2400 years ago the Athenians offered their capacity of assessing risks before making decisions. From the Pericle's Funeral Oration in Thurcydidas' "History of the Peloponnesian War" (started in 431 B.C.), we can read:

> We Athenians in our persons, take our decisions on policy and submit them to proper discussion. The worst thing is to rush into action before consequences have been properly debated. And this is another point where we differ from other people. We are capable at the same time of taking risks and assessing them beforehand. Others are brave out of ignorance; and when they stop to think, they begin to fear. But the man who can most truly be accounted brave is he who best knows the meaning of what is sweet in life, and what is terrible, and he then goes out undeterred to meet what is to come.

But the Greeks did not develop a quantitative approach to risk. They had no numbers, and without numbers there are no odds and probabilities. And without odds and probabilities, the natural way of dealing with risk is to appeal to the gods and the fates; risk is wholly a matter of gut. These are words in the spirit of Peter Bernstein in *Against the Gods* (1996), who describes in a fascinating way how our understanding of risk has developed over centuries. Until the theory of probability was sufficiently developed, our ability to define and manage risk was necessarily limited. Bernstein asks rhetorically, What distinguishes the thousands of years of history from what we think of as modern times? The past has been full of brilliant scientists, mathematicians, investors, technologists, and political philosophers, whose achievements

---

Foundations of Risk Analysis T. Aven
© 2003 John Wiley & Sons, Ltd ISBN: 0-471-49548-4

were astonishing; think of the early astronomers or the builders of the pyramids. The answer Bernstein presents is the mastery of risk; the notion that the future is more than a whim of the gods and that men and women are not passive before nature. By understanding risk, measuring it and weighing its consequences, risk-taking has been converted into one of the prime catalysts that drives modern Western society. The transformation in attitudes towards risk management has channelled the human passion for games and wagering into economic growth, improved quality of life, and technological progress. The nature of risk and the art and science of choice lie at the core of our modern market economy that nations around the world are hastening to join.

Bernstein points to the dramatic change that has taken place in the last centuries. In the old days, the tools of farming, manufacturing, business management, and communication were simple. Breakdowns were frequent, but repairs could be made without calling the plumber, the electrician, the computer scientist – or the accountants and the investment advisers. Failure in one area seldom had direct impact on another. Today the tools we use are complex, and breakdowns can be catastrophic, with far-reaching consequences. We must be constantly aware of the likelihood of malfunctions and errors. Without some form of risk management, engineers could never have designed the great bridges that span the widest rivers, homes would still be heated by fireplaces or parlour stoves, electric power utilities would not exist, polio would still be maiming children, no airplanes would fly, and space travel would be just a dream.

Traditionally, hazardous activities were designed and operated by references to codes, standards and hardware requirements. Now the trend is a more functional orientation, in which the focus is on what to achieve, rather than the solution required. The ability to address risk is a key element in such a functional system; we need to identify and categorize risk to provide decision support concerning choice of arrangements and measures.

The ability to define what may happen in the future, assess associated risks and uncertainties, and to choose among alternatives lies at the heart of the risk management system, which guides us over a vast range of decision-making, from allocating wealth to safeguarding public health, from waging war to planning a family, from paying insurance premiums to wearing a seat belt, from planting corn to marketing cornflakes.

To be somewhat more detailed, suppose an oil company has to choose between two types of concept, A and B, for the development of an oil and gas field. To support the decision-making, the company evaluates the concepts with respect to a number of factors:

- *Investment costs*: there are large uncertainties associated with the investment costs for both alternatives. These uncertainties might relate to the optimization potential associated with, among other things, reduction in management and engineering man-hours, reduction in fabrication costs and process plant optimization. The two alternatives are quite different with respect to cost reduction potential.

- *Operational costs*: there is greater uncertainty in the operational cost for B than for A as there is less experience with the use of this type of concept.
- *Schedules*: the schedule for A is tighter than for B. For A there is a significant uncertainty of not meeting the planned production start. The cost effect of delayed income and back-up solutions is considerable.
- *Market deliveries and regularity*: the market has set a gas delivery (regularity) requirement of 99%, i.e. deliveries being 99% relative to the demanded volume. There are uncertainties related to whether the alternatives can meet this requirement, or in other words, what the cost will be to obtain sufficient deliveries.
- *Technology development*: alternative A is risk-exposed in connection with subsea welding at deep water depth. A welding system has to be developed to meet a requirement of approximately 100% robotic functionality as the welding must be performed using unmanned operations.
- *Reservoir recovery*: there is no major difference between the alternatives on reservoir recovery.
- *Environmental aspects*: alternative B has the greater potential for improvement with respect to environmental gain. New technology is under development to reduce emissions during loading and offloading. Further, the emissions from power generation can be reduced by optimization. Otherwise the two concepts are quite similar with respect to environmental aspects.
- *Safety aspects*: for both alternatives there are accident risks associated with the activity. There seems to be a higher accident risk for A than for B.
- *External factors*: concept A is considered to be somewhat advantageous relative to concept B as regards employment, as a large part of the deliveries will be made by the national industry.

Based on evaluations of these factors, qualitative and quantitative, a concept will be chosen. The best alternative is deemed to be the one giving highest profitability, no fatal accidents and no environmental damage. But it is impossible to know with certainty which alternative is the best as there are risks and uncertainties involved. So the decision of choosing a specific alternative has to be based on predictions of costs and other key performance measures, and assessments of risk and uncertainties. Yet, we believe, and it is essentially what Bernstein tells us, that such a process of decision-making and risk-taking provides us with positive outcomes when looking at the society as a whole, the company as a whole, over a certain period of time. We cannot avoid 'negative' outcomes from time to time, but we should see 'positive' outcomes as the overall picture.

As a second example, let us look at a stock market investor. At a particular moment, the investor has x million dollars with which to buy stocks. To simplify, say that he considers just three alternatives: A, B and C. What stocks should he buy? The decision is not so simple because there are risks and uncertainties involved. As support for his decision, he analyses the relevant companies. He would like to know more about how they have performed so far, what their goals and strategies are, what makes them able to meet these goals and strategies, how vulnerable the companies are with respect to key personnel, etc. He would also analyse the industries the companies belong to. These analyses give insight into the risks and uncertainties, and they provide a basis for the decision-making. When the investor makes his choice, he believes he has made the right choice, but only time will tell.

As a final example, let us consider a team of doctors that consider two possible treatments, A and B, for a patient who has a specific disease. Treatment A is a more comprehensive treatment, it is quite new and there are relatively large uncertainties about how it will work. There are some indications that this treatment can give very positive results. Treatment B is a more conventional approach, it is well proven but gives rather poor results. Now, which treatment should be chosen? Well, to make a decision, risks and uncertainties first have to be addressed. The team of doctors have thoroughly analysed these risks and uncertainties, and to some extent reduced them. For the patient it is important to hear the doctors' judgements about his chances of being cured and about the possible side effects of the treatments. Then the patient makes his decision.

More examples will be presented in the coming chapters.

1.2 THE NEED TO DEVELOP A PROPER RISK ANALYSIS FRAMEWORK

Bernstein's concludes that the mastery of risk is a critical step in the development of modern society. One can discuss the validity of his conclusion, but there should be no doubt that risk and uncertainty are important concepts to address for supporting decision-making in many situations. The challenge is to know how do describe, measure and communicate risk and uncertainty. There is no clear answer to this. We cannot find an authoritative way of approaching risk and uncertainty. We do need one. We all have a feel of what risk means, but if we were asked to measure it, there would be little consensus. The word 'risk' derives from the early Italian *risicare*, which means 'to dare'. Webster's Dictionary (1989) has several definitions of 'risk'; here are some of them:

- expose to the chance of injury or loss;
- a hazard or dangerous chance;
- the hazard or chance of loss;
- the degree of probability of such loss.

We are not yet ready to define what we mean by risk in this book, but the definition in Chapter 3 is closely related to uncertainty, a concept that is equally difficult to define as risk. Webster's Dictionary refers among other things, to the following definitions of 'uncertainty':

- not definitely ascertainable or fixed;
- not confident;
- not clearly or precisely defined;

- vague, indistinct;
- subject to change, variable;
- lack of predictability.

The ambiguity surrounding the notions of risk and uncertainty is also reflected in the way the different applications and disciplines approach risk and uncertainty. This will become apparent in Chapter 2, which reviews some common thinking about risk in different applications and disciplines.

The terminology and methods used for dealing with risk and uncertainty vary a lot, making it difficult to communicate across different applications and disciplines. We also see a lot of confusion about what risk is and what should be the basic thinking when analysing risk and uncertainty within the various applications. This is not surprising when we look at the risk literature, and the review in the next chapter will give some idea of the problems. Reference is made to so-called classical methods and Bayesian methods, but most people find it difficult to distinguish between the alternative frameworks for analysing risk. There is a lack of knowledge about what the analyses express and the meaning of uncertainty in the results of the analyses, even among experienced risk analysts. The consequence of this is that risks are often very poorly presented and communicated.

Nowadays there is an enormous public concern about many aspects of risk. Scientific advances, the growth in communications and the availability of information have led to stronger public awareness. Few risks are straightforward; there are competing risks to balance, there are trade-offs to make and the impacts may be felt across many sections of society and the environment. Science, medicine and technology can help us to understand and manage the risks to some extent, but in most cases the tasks belong to all of us, to our governments and to public bodies. Therefore we need to understand the issues and facilitate communication among all parties concerned. The present nomenclature and tools for dealing with risk and uncertainty are confusing and do not provide a good framework for communication.

Furthermore, aspects of society with inherent risk and uncertainty have changed in recent years. This applies, among other things, to complex technology with increased vulnerability, information and communication technology, biotechnology and sabotage. People require higher safety and reliability, and environmental groups have intensified their activities. The societal debate related to these issues is characterized by people talking at cross purposes, by mistrust as objective facts are mixed with judgements and values, and the cases are often presented in a non-systematic way as far as risk and uncertainty are concerned. More than ever there is a need for decision-support tools addressing risk and uncertainty.

It is our view that the concepts of risk and risk analysis have not yet been sufficiently developed to meet the many challenges. A common approach is needed that can give a unifying set-up for dealing with risk and uncertainty over the many applications. It is necessary to clarify what should be the basis of risk analysis. We search for a common structure, and philosophy, not a straitjacket. Business needs a different set of methods, procedures and models than for example medicine. But there is no reason why these areas should have completely different perspectives on how to think when approaching risk and uncertainty, when the basic problem is the same – to reflect our knowledge and lack of knowledge about the world.

This book presents such a unifying approach, which we believe will meet the many challenges and help to clarify what should be the definition of risk and the basis of risk analysis. To deal with risks related to the profit from one or several investment projects or stocks, production loss and occurrence of accidental events, it is essential that economists, finance analysts, project managers, safety and production engineers are able to communicate. Currently this communication is difficult. The typical approaches to risk and risk analysis adopted in engineering and in business and project management represent completely different views, making the exchange of ideas and results complicated and not very effective. In traditional engineering applications, risk is a physical property to be analysed and estimated in the risk analysis, the quantitative risk analysis (QRA) and the probabilistic safety analysis (PSA); whereas in business and project management, risk is seen more as a subjective measure of uncertainty.

We need to rewrite the rules of risk and risk analysis. And our starting point is a review of the prevailing thinking about risk in different applications and disciplines.

BIBLIOGRAPHIC NOTES

The literature covers a vast number of papers and books addressing risk and uncertainty. Many provide interesting examples of real-life situations where risk and uncertainty need to be analysed and managed. Out of this literature we draw attention to Clemen (1996), Moore (1983), Hertz and Thomas (1983), and Koller (1999a, 1999b), as these books are closely linked to the main applications that we cover in this book.

The challenges related to description, measurement and communication of risk and uncertainty have been addressed by many researchers. They will be further discussed in Chapter 2, and more bibliographic notes can be found there.

2

Common Thinking about Risk and Risk Analysis

In this chapter we review some main lines of thinking about risk and risk analysis, focusing on industry and business. The purpose is not to give a complete overview of the existing theory, but to introduce the reader to common concepts, models and methods. The exposition highlights basic ideas and results, and it provides a starting point for the theory presented in Chapters 3 to 5. First we look into accident risk, mainly from an industry view point. We cover accident statistics, risk analysis and reliability analysis. Then we consider economic risk, focusing on business risk. Finally we discuss the ideas and methods we have reviewed and draw some conclusions.

2.1 ACCIDENT RISK

2.1.1 Accident Statistics

To many people, risk is closely related to accident statistics. Numerous reports and tables are produced showing the number of fatalities and injuries as a result of accidents. The statistics may cover the total number of accidents associated with an activity within different consequence categories (loss of life, personal injuries, material losses, etc.) and they could be related to different types of accident, such as industrial accidents and transport accidents. Often the statistics are related to time periods, and then time trends can be identified. More detailed information is also available in some cases, related to, for example, occupation, sex, age, operations, type of injury, etc.

Do these data provide information about the future, about risk? Yes, although the data are historical data, they would usually provide a good picture of what to expect in the future. If the numbers of accidental deaths in traffic during the previous five years are 1000, 800, 700, 800, 750, we know a lot about risk, even though we have not explicitly expressed it by formulating predictions and uncertainties. This is risk related to the total activity, not to individuals. Depending on your driving habits, these records could be more or less representative for you.

Accident statistics are used by industry. They are seen as an essential tool for management to obtain regular updates on the number of injuries (suitably defined) per hour of working, or any other relevant reference, for the total company and divided into relevant organizational units. These numbers provide useful information about the safety and risk level within the relevant units. The data are historical data, but assuming a future performance of systems and human beings along the same lines as this history, they give reasonable estimates and predictions for the future.

According to the literature, accident statistics can be used in several ways:

- to monitor the risk and safety level;
- to give input to risk analyses;
- to identify hazards;
- to analyse accident causes;
- to evaluate the effect of risk reducing measures;
- to compare alternative area of efforts and measures.

Yes, we have seen accident statistics used effectively in all these ways, but we have also seen many examples of poor use and misuse. There are many pitfalls when dealing with accident statistics, and the ambitions for the statistics are often higher than is achieved. In practice it is not so easy to obtain an effective use of accident statistics. One main challenge is interpreting historical data to estimate future risks. Changes may have occurred so that the situation now being analysed is quite different from the situation the data were based on, and the amount of data could be too small for making good predictions.

Suppose that we have observed 2 and 4 accidents leading to injuries (suitably defined) in a company in two consecutive years. These numbers give valuable information about what has happened in these two years, but what do they say about risk? What do the numbers say about the future? For the coming year, should we expect 3 accidents leading to injuries, or should we interpret the numbers such that it is likely that 4 or more accidents would occur. The numbers alone do not provide us with one unique answer. If we assume, as a thought experiment, that the performance during the coming years is as good (bad) as in previous years, then we would see 3 accidents per year on the average. If we see a negative trend, we would indicate 4 accidents per year, or even a higher number. But what about randomness, i.e. variations that are not due to a systematic worsening or improvement of the safety level? Even if we say that 3 events would occur on the average per year, we should expect that randomness could give a higher or lower number next year. A common model to express event streams such as accidents is the Poisson model. If we use this model and assume 3 events to occur on the average, the probabilities of 0 events and 1 event during one year are equal to 5% and 15%, respectively. The probability of 5 or more events is 20%; for 6 and 7 the corresponding probabilities are 8% and 3%. So even if 5 events occur, we should be careful in concluding that the safety level has been significantly decreased – the increase in accidental events could be a result of randomness. At a level of 7 events or more, we will be reasonably sure if we assert that a worsening has occurred, because in this case there is not more than a probability of 3% of concluding that the safety level has decreased when this is not the case.

Our reasoning here is similar to classical statistical hypothesis testing, which is commonly used for analysing accident data. The starting point is a null hypothesis (3 events on the average per year) and we test this against a significant worsening (improvement) of the accident rate. We require a small probability (about 5–10%) for rejecting the null hypothesis when the null hypothesis is true, i.e. make an erroneous rejection of the null hypothesis. This is a basic principle of classical statistical thinking. The problem with this principle is that the data must give a very strong message before we can conclude whether the safety level has worsened (improved). We need a substantial amount of data to enable the tests to reveal changes in the safety level. Seven or more events give support for the conclusion that the safety level has worsened, and this will send a message to management about the need for risk-reducing measures.

Note that the statistical analysis does not reveal the causes of the decrease in safety level. More detailed analysis with categorized data is required to identify possible causes. However, the number of events in each category would then be small, and inference would not be very effective.

Trend analyses are seen as a key statistical tool for identifying possible worsening or improvement in the safety level. The purpose of a trend analysis is to investigate whether trends are present in the data, i.e. whether the data show an increase or decrease over time that is not due to randomness. Suppose we have the observations given in Table 2.1. We assume that the number of working hours is constant for the time period considered. The question now is whether the data show that a trend is present, i.e. a worsening in the safety level that is not due to randomness. And if we can conclude there is a trend, what are its causes? Answering these questions will provide a basis for identifying risk-reducing measures that can reverse the trend.

Statistical theory contains a number of tests to reveal possible trends. The null hypothesis in such tests is no trend. It requires a considerable amount of data and a strong tendency in the data in order to give rejection of this null hypothesis. In Table 2.1, we can observe that there is some tendency of an increasing number of injuries as a function of time, but a statistical test would not prove that we have a significant increase in injuries. The amount of data

Table 2.1 Number of injuries

Month	1	2	3	4	5	6
Number of injuries	1	2	1	3	3	5

is too small – the tendency could be a result of randomness. To reject the null hypothesis a large change in the number of injuries would be required, but hopefully such a development would have been stopped long before the test gives the alarm.

To increase the amount of data, we may include data of near misses and deviations from established procedures. Such events can give a relatively good picture of where accidents might occur, but they do not necessarily give a good basis for quantifying risk. An increase in the number of near misses could be a result of a worsening of the safety, but it could also be a result of increased reporting.

We conclude that in an active safety management regime, classical statistical methods cannot be used as an isolated instrument for analysing trends. We must include other information and knowledge besides the historical data. Based on their competence and position, someone must transform the data to a view related to the possible losses and damages, where consideration is given to uncertainties and randomness. Information from near-miss reporting is one aspect, and another aspect is insight into the relevance of the data for describing future activities.

When the data show a negative trend as in Table 2.1 above, we should conclude immediately that a trend is present – the number of events is increasing. We can observe this without any test. Quick response is required as any injury is unwanted. We should not explain the increase by randomness. And more detailed statistical analysis is not required to conclude this. Then we need to question why this trend is observed and what we can do to reduce the number of injuries. We need some statistical competence, but equally as important, or perhaps even more important, is the ability to find out what can cause injuries, how hazardous situations occur and develop into accidents, how possible measures can reduce risk, etc. After having analysed the different accidental events, seen in relation to other relevant information and knowledge, we need to identify the main factors causing this trend, to the best of our ability. This will imply more or less strong statements depending on the confidence we have about the causes. Uncertainty will always be present, and sometimes it will be difficult to identify specific causes. But this does not mean that the accidental events are due to randomness. We do not know. This would be the appropriate conclusion here.

Statistical testing should be seen more as a screening instrument for identifying where to concentrate the follow-up when studying several types of accidental event. Suppose we have to look into data of more than 100 hazards. Then some kind of identification of the most surprising results would be useful, and statistical testing could be used for this purpose.

A basic requirement is that historical data are correct – they are reliable. In our injuries example it would be difficult in many cases to make accurate measurements. Psychological and organizational factors could result in underreporting. We may think of an organizational incentive structure where absence of injuries is rewarded. Then we may find that some injuries are not reported as the incentive structure is interpreted as 'absence of reported injuries'. So judgements are required – we cannot base our conclusions on the data alone.

Another measurement problem is related to the specification of relevant reference or normalizing factors to obtain suitable accident or failure rates, for example the number of working hours, opportunities of failure, and so on.

Historical data on a certain type of accident, for example an injury rate, provide information about the safety level. But we cannot use just one indicator, such as the injury rate, to draw conclusions about development in the safety level as a whole. The safety level is more than the number of injuries. A statement concerning the safety level based on observations of the injury rate only, would mostly have low validity.

Most researchers and analysts seem to consider statistical testing as a strongly scientific approach as it can make objective assessments on the probabilities of making errors as well as the probability of correctly rejecting the null hypothesis. Probability is defined according to the relative frequency interpretation, meaning that probability is an objective quantity expressing the long-run fraction of successes if the experiment were repeated for real or hypothetically an infinite number of times. Furthermore it is assumed that the data (here the number of accidents) follow some known probability law, for example the Poisson distribution or the normal (Gaussian) distribution. The problem is that these probabilities and probability models cannot be observed or verified – they are abstract theoretical quantities based on strong assumptions. Within its defined framework the tool is precise, but precision is not interesting if the framework conditions are inappropriate.

In the case of accidents with severe damage and losses, the amount of data would normally be quite limited, and the data would give a rather poor basis for predicting the future. For example, in a company there would normally be few fatal accidents, so a report on fatalities would not be so useful for expressing risk, and it would be difficult to identify critical risk factors and study the effect of risk-reducing measures. Even with large amounts of accident data it is not clear that fatality reports are useful for expressing risk. What we need is a risk analysis.

2.1.2 Risk Analysis

We consider an offshore installation producing oil and gas. As part of a risk analysis on the installation, a separate study is to investigate the risk associated with the operation of the control room that is placed in a compressor module. Two persons operate the control room. The purpose of the study is to assess risk to the operators as a result of possible fires and explosions in the module and to evaluate the effect of implementing risk-reducing measures. Based on the study a decision will be made on whether to move the control out of the module or to implement some other risk-reducing measures. The risk is currently considered to be too high, but the management is not sure what is the overall best arrangement taking into account both safety and economy.

We will examine this control room study by focusing on the following questions:

- How is risk expressed?
- What is the meaning of probability and risk?

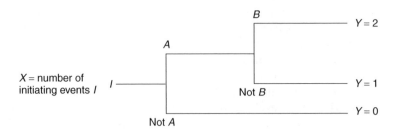

Figure 2.1 Event tree example

- How is uncertainty understood and addressed?
- What is the meaning of a model?
- How do we use and understand parametric probability models like the Poisson model?

We will assume that the study is simply based on one event tree as shown in Figure 2.1. The tree models the possible occurrence of gas leakages in the compression module during a period of time, say one year. A gas leakage is referred to as an initiating event. The number of gas leakages is denoted by X. If an initiating event I occurs, it leads to Y fatalities, where $Y = 2$ if the events A and B occur, $Y = 1$ if the events A and not B occur, and $Y = 0$ if the event A does not occur. We may think of the event A as representing ignition of the gas and B as explosion.

Now, what would a risk analyst do, following today's typical industry practice? There are many different answers; we will look at two, a fairly simple approach and a more sophisticated approach.

Best-estimate approach

The simple approach, here called the best-estimate approach, goes like this. First the frequency of leakages and of the probabilities of ignition and explosion are estimated. Then the frequency of events resulting in 2 and 1 fatalities are calculated by multiplying these estimates. The probability of having two or more accidents with fatalities during one year is ignored. If for example a frequency of 1 leakage per year is estimated, and an ignition probability of 0.005 and an explosion probability of 0.1, then an estimate of 0.0005 events resulting in 2 fatalities per year is derived, and an estimate of 0.0045 events resulting in 1 fatality per year. Combining these numbers, the PLL (potential loss of lives) and FAR (fatal accident rate) values can be calculated. The PLL value represents the average number of fatalities per year and is equal to $0.0045 \times 1 + 0.0005 \times 2 = 0.0055$, and the FAR value represents the average number of fatalities per 100 million exposed hours and is equal to $[0.0055/2 \times 8760] \times 10^8 = 31$, assuming there are two persons at risk at any time, so that the total hours of risk exposure is equal to 2×8760 per year.

To estimate the leakage frequency, ignition probability and explosion probability, observations from similar activities (often known as hard data) and judgements are used. Detailed modelling of the ignition probability may be carried out in some cases. This modelling covers the probability of exposure to flammable mixtures accounting for release characteristics (e.g. duration, flow) and the dispersion or spreading of the gas (e.g. geometry, ventilation) in the module, as well as characteristics of potential ignition sources, for example electrical equipment and hot work. The modelling makes it possible to study the influence on risk of mitigation measures (e.g. shutdown, working procedures) and is expected to give more accurate estimates of the ignition probability.

These risk numbers are presented to management along with typical FAR values for other activities. Changes in the risk estimates are also presented to show what happens when possible risk-reducing measures are incorporated.

In practice, analysts also focus on other risk indices, for example the probability of a safety function impairment during a specific year. An example of a safety function is: People outside the immediate vicinity of an accident shall not be cut of from all escape routes to a safe area.

Now, what do these estimates express and what about uncertainties? If these questions are put forward, we will receive a variety of answers. Here is a typical answer:

> The results of any risk analysis are inevitably uncertain to some degree. The results are intended to be 'cautious best estimates'. This means that they attempt to estimate the risks as accurately as possible, but are deliberately conservative (i.e. tending to overestimate the risks) where the uncertainties are largest. Because of the inevitable limitations of the risk analysis approach, it must be acknowledged that the true risks could be higher or lower than estimated.
>
> These uncertainties are often considered to amount to as much as a factor of 10 in either direction. A detailed analysis of the confidence limits on the results would be prohibitively complex, and in itself extremely uncertain.

We do not find this satisfactory. The approach is in fact not complete, as it does not seriously deal with uncertainty. To explain our view in more detail, we will formalize the above presentation of the 'best-estimate' approach.

In this framework, risk is supposed to be an objective characteristic or property of the activity being analysed, expressed by probabilities and statistically expected values of random variables such as the number of fatalities Y. To be more specific, in the above example we draw attention to $P(Y = 2)$ and EY. We may think of this probability as the long-run proportion of observations having events with two fatalities when considering (hypothetically) an infinite number of similar installations, and the expected value as the mean number of fatalities when considering (hypothetically) an infinite number of similar installations. This true risk is estimated in the risk analysis, as demonstrated in the above example. Note that the risk analyst above has estimated $P(Y = 2)$ by estimating the expected number of leakages leading to two fatalities. These underlying probabilistic quantities are approximately equal in this case as the expected number of leakages resulting in two fatalities during a period of one year is about the same as the probability of having one leakage resulting in two fatalities during one year. The probability of having two or more leakage scenarios with fatalities is negligible compared to having one.

So the risk analyst is providing estimates of the true risk, i.e. the probabilities and expected values. The PLL value is defined as the expected number of fatalities per year, and 0.0055 is an estimate of this value. The interpretation is mentioned above; it is the average number of fatalities per year when considering an infinite number of similar installations. The FAR value is defined as the expected number of fatalities per 100 million exposed hours.

We refer to this framework as the *classical approach* to risk analysis. Assuming that all input data to the event tree model are observed data (hard data), the approach is consistent with traditional statistical modeling and analysis as described in most textbooks in statistics. Risk is a function of unknown parameters to be estimated. Using statistical principles and methods, estimates are derived for the parameters, and this gives the estimates of the relevant risk indices. Let r represent such a risk index, and let f be a model linking r and some parameters $\mathbf{q} = (q_1, q_2, \ldots, q_v)$ on a more detailed level. Thus we can write

$$r = f(\mathbf{q}). \tag{2.1}$$

In the above example, r may be equal to $P(Y = 2)$ or EY, $\mathbf{q} = (EX, P(A), P(B|A))$ and f equals the event tree model based on the assumption that the probability of having two or more events leading to fatalities during one year is ignored. This model expresses, for example, that

$$P(Y = 2) = EX \cdot P(A) \cdot P(B|A). \tag{2.2}$$

In the classical approach, we estimate the parameters $\mathbf{q}$, and through the model f we obtain an estimate of r. Replacing $\mathbf{q}$ by estimates $\widehat{\mathbf{q}}$, we can write

$$\widehat{r} = f(\widehat{\mathbf{q}}).$$

In this set-up there exist true values of $\mathbf{q}$ and r, but as f is a model, i.e. a simplification of the real world, equation (2.1) is not necessarily correct for the true values of $\mathbf{q}$ and r. Thus there are two main contributors to uncertainty in $\widehat{r}$'s ability to estimate r: the estimates $\widehat{\mathbf{q}}$ and the choice of model f. There is, however, no formal treatment of uncertainty in the best-estimate approach.

The main features of the classical approach, focusing on best estimates, are summarized in Figure 2.2. Note that in a classical setting the probabilities are considered elements of the world (the reality), properties of the physical world like height and weight. A drawing pin, for example, has a weight and a probability, p, of landing with its point in the air. To determine or estimate the weight and the probability, we perform measurements. For probabilities, repeated experiments are required. Throwing the drawing pin over and over

COMMON THINKING ABOUT RISK AND RISK ANALYSIS

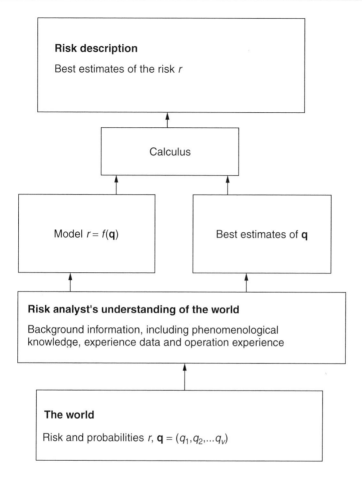

Figure 2.2 Basic elements of a risk analysis. Classical approach based on best estimates

again, we are able to accurately estimate p by observing the proportion of times the pin lands with its points in the air. This is the classical view; we will discuss this way of thinking in Section 2.3.1.

Here are the main steps of the risk analysis when this approach is adopted:

1. Identify suitable risk indices.
2. Develop a model of the activity or system being analysed, linking more detailed elements of the system and the overall risk indices.
3. Estimate unknown parameters of the model.
4. Use the model to generate an estimate of the risk indices.

Risk estimates obtained by models are sometimes known as notional risk, in contrast to actuarial risk, which is based on hard data only (Vinnem 1999).

Classical approach including uncertainty analysis

In the classical approach presented above, we identified the two main contributors to uncertainty as the parameter estimates $\widehat{\mathbf{q}}$ and the choice of model f. The model uncertainty could be a result of:

- Faulty or insufficient system or activity definition. This is mainly a problem in the earliest phases of a project when there will be limited information about technical solutions, operation and maintenance philosophies, logistic conditions, etc.
- Limitations and errors in the model itself. The analyst could have omitted some important risk contributors, the model could be extremely inaccurate, etc. This item also includes simplifications to reduce computing time, e.g. using only four wind directions and strengths to represent an infinite number of combinations in the gas dispersion calculations.

The uncertainty related to the input parameters $\widehat{\mathbf{q}}$ could be a result of:

- Data are used which are not representative for the actual equipment or event, the data are collected from non-representative operating and environmental conditions, etc.
- The data analysis methods producing the estimates are not adequate.
- Wrong information, perhaps concerning the description of the equipment.
- Insufficient information, perhaps concerning how to use the equipment.
- Statistical variation, the data basis is small.

By using quantities like variance, standard deviation and confidence interval, it is possible to express the statistical variation based on observed data. For many risk analysts this is seen as the proper way of dealing with uncertainty, and confidence intervals are quite often presented for some of the initiating events, for example related to leakages. Suppose we have observed 2, 1, 0, 1, 0, 1, 0, 0, 0, 2, 3, 2 leakages from similar activities. Based on this we find a mean of 1 per year, which we use as the estimate for the future leakage occurrence rate, $\lambda = EX$. Assuming that the number of leakages follows a Poisson process with rate λ (see Appendix A, p. 165), we can compute a confidence interval for λ. A 90% confidence interval is given by (0.58, 1.62). The details are presented in Appendix A, p. 168. Note that a confidence interval is based on hard data and the classical relative frequency interpretation of probability. When the interval is calculated, it will either include the true value of λ or it will not. If the experiment were repeated many times, the interval would cover the true value of λ 90% of the time. Thus we would have a strong confidence that λ is covered by (0.58, 1.62), but it is wrong to say that there is a 90% probability that λ is included in this interval. The parameter λ is not stochastic. It has a true but unknown value.

It is, however, difficult to quantify other sources of uncertainty than the statistical variation. Consequently, the uncertainty treatment is rather incomplete.

A possible emphasis on statistical variation leads to a rather inadequate picture of the overall uncertainty of estimates.

Other approaches for dealing with uncertainty of the risk and its estimate are therefore needed. The simplest approach seen in practice normally gives very wide intervals, but it is not so difficult to carry out. The idea is to identify the extreme values of the parameters of the model. The greatest possible variations (most conservative and most optimistic) in the input data are determined. For practical reasons, not all uncertainties attached to every input are included. The main areas of uncertainty included in the analysis are identified using experience and judgement. The effects of the modelled variations on the risks are then calculated for two cases: a most pessimistic case, where all model variations which tend to increase the risk are assumed to act together, and a most optimistic case, where all modelled variations which tend to decrease the risk are assumed to act together. The range between the two cases indicates the uncertainty of the risk and the best estimate of the risk. Analysts using this approach link it to confidence intervals, but acknowledge that they are not really the same. We know that they are in fact not related at all. A confidence interval expresses statistical variation, whereas the extreme values approach produces intervals reflecting all types of uncertainties associated with the parameters of the model, and these intervals are based on subjective evaluations.

For our numerical example, we determine a most pessimistic leakage frequency of 2 per year and a most optimistic one as 0.5. For the ignition probability the corresponding values are 0.01 and 0.001, and for the explosion probability 0.2 and 0.05. This gives an interval of [0.0005, 0.024] for the PLL and an interval of [3, 137] for the FAR value. We see that the intervals produced are very wide, as expected since the calculations are based on maximum and minimum values for all parameters.

A more precise approach has been developed, and it is a common way of dealing with uncertainty in risk analyses. When we speak about the classical approach including uncertainty analysis, it is this more precise approach that we have in mind.

The uncertainty problem of risk analysis is solved by dividing uncertainty into two categories: the stochastic (aleatory) uncertainty and the knowledge-based (epistemic) uncertainty. The aleatory uncertainty stems from variability in known (or observable) populations and represents randomness in samples, whereas the epistemic uncertainty comes from lack of basic knowledge about fundamental phenomena. Probability is used as a measure of uncertainty in both cases, but the interpretation is different. To make this difference more precise, let us consider our offshore installation example. The stochastic uncertainties are represented by the random variable X, the number of leakages; A, the event that the gas is ignited; B, the event that explosion occurs; and the number of fatalities Y. The random variable X is assumed to follow a Poisson distribution with mean λ, meaning that the number of leakages has a variation according to this distribution when considering an infinite population of similar installation years. In practice, 'infinite' is interpreted as large or very large. Similarly, we use a relative frequency to quantify the variations related to ignition or not ignition, and explosion or not explosion. For example, $P(A)$ represents the proportion of leakages resulting in ignition when considering an infinite number of similar situations. Having introduced these measures of aleatory uncertainty, it remains to describe the epistemic uncertainty related to the true values of λ, $P(A)$ and $P(B|A)$. This is done by expressing subjective probabilities for these quantities. Let us look at a simple numerical example. For λ the analyst allows for three possible values: 0.5, 1 and 2. The analyst expresses his degree of belief with respect to which value is the true one by using the corresponding probabilities 0.25, 0.50 and 0.25. So the analyst has the strongest belief in λ equalling 1, but he also has rather strong belief in λ equalling 0.5 or 2. For the probabilities $P(A)$ and $P(B|A)$ he also considers three values, 0.001, 0.005, 0.01 and 0.05, 0.1, 0.2 respectively, with corresponding probabilities 0.25, 0.50 and 0.25 in both cases. These numbers are supposed to be based on all relevant information, hard data and engineering judgements. From these probabilities we can calculate an epistemic uncertainty distribution over $P(Y = y)$, $y = 0, 1, 2$. For notational convenience, let us write p_y for $P(Y = y)$. To illustrate the calculations, consider the highest value of p_2, i.e. $p_2 = 2 \times 0.01 \times 0.2 = 0.004$. Then we obtain

$$P(p_2 = 0.004) = 0.25 \times 0.25 \times 0.25 = 0.0156.$$

The complete uncertainty distributions are presented in Tables 2.2 and 2.3. From the uncertainty distributions we can compute so-called credibility intervals. For example, [4,120] is approximately a 90% credibility interval for the FAR value, meaning that our probability is 90% that the true FAR value is included in the interval.

It is common to establish uncertainty distributions by the use of Monte Carlo simulation. The basic idea of Monte Carlo simulation is to use a computer random number generator to generate realizations of the system performance by drawing numbers from the input probability distributions. For our example the computer draws numbers from the distributions for λ, and $P(A)$ and $P(B|A)$.

Table 2.2 Uncertainty distribution for p_2, $p_1 + p_2$ and the PLL value

Risk index	Value of risk index					
	≤ 0.001	(0.001– 0.002]	(0.002– 0.004]	(0.004– 0.01]	(0.01– 0.02]	(0.02– 0.032]
p_2	0.89	0.09	0.02	0.00	0.00	0.00
$p_1 + p_2$	0.19	0.06	0.13	0.56	0.00	0.06
PLL	0.06	0.13	0.19	0.31	0.25	0.06

Table 2.3 Uncertainty distribution for the true FAR value

FAR	≤ 10	(10–20]	(20–30]	(30–40]	(40–50]	(50–75]	(75–100]	(100–150]
Prob.	0.19	0.19	0.08	0.23	0.0	0.25	0.00	0.06

For the numbers drawn for λ, and $P(A)$ and $P(B|A)$, we compute the corresponding value of p_y using the event tree model, i.e. an equation like (2.1). This procedure is repeated many times, and with a sufficient number of repetitions we will be able to determine the same value of the uncertainty distribution $H_y(p) = P(p_y \le p)$, as done by the analytical calculations.

To represent the complete uncertainty distributions, we use summarizing measures such as the mean and the variance. The mean is of particular interest. In our example it follows from the model structure (2.2) that the means of the uncertainty distributions are equal to the risk measures with the mean values used as parameters. To see this, note that the risk measure p_2 is equal to $q_1 q_2 q_3$, where $q_1 = \lambda$, $q_2 = P(A)$ and $q_3 = P(B|A)$. Then using independence in the assessment of the uncertainties of the q_i, and applying the rules for computing expectations and probabilities by conditioning, we obtain

$$\begin{aligned} Ep_2 &= E[q_1 q_2 q_3] \\ &= E[q_1]E[q_2]E[q_3] \\ &= E[E(X|q_1)]E[P(A|q_2)]E[P(B|q_3, A)] \\ &= EX \cdot P(A) \cdot P(B|A). \end{aligned}$$

In other words, the mean of the uncertainty distribution is equal to the related risk measure with the mean values used as parameters. This result does not hold in general. The mean of the uncertainty distribution is referred to as the predictive distribution of Y. We have $P(Y = i) = Ep_i$, hence the predictive distribution is a measure of both the aleatory and the epistemic uncertainty; the aleatory uncertainty is expressed by p_i and the epistemic uncertainty is expressed by the uncertainty in the true value of p_i. The predictive distribution provides a tool for prediction of Y reflecting these uncertainties. Note that the predictive distribution is not a total measure of uncertainty, as it does not reflect uncertainty related to the choice of the model f. The predictive distribution can be seen as an estimate of the true value of the risk index p_i, as it is equal to the mean of the uncertainty distribution. Of course, the mean could give a more or less good picture of this distribution.

Using a more general set-up, the predictive distribution is given by

$$Er = Ef(\mathbf{q}),$$

where the expectation is with respect to the epistemic uncertainty of the parameters $\mathbf{q}$ of the model f. In many applications, such as the one considered here, the function f is linear in each argument, and we obtain $Ef(\mathbf{q}) = f(E\mathbf{q})$, where $E\mathbf{q} = (Eq_1, Eq_2, \ldots, Eq_v)$. Thus

$$Er = f(E\mathbf{q}).$$

So if r is the true value of $P(D)$ for some event D, a measure of uncertainty of D covering stochastic and epistemic uncertainty is in this case given by $P(D) = f(E\mathbf{q})$.

The above classical approaches introduce two levels of uncertainty: the value of the observable quantities and the correct value of the risk. The result is often that both the analysis and the results of the analysis are considered uncertain. This does not provide a good basis for communication and decision-making. In the above example we derived a 90% credibility interval for the FAR value of [4,120]. In larger and more complete analyses, we would obtain even wider intervals. What is then the message from the analysis? We have a best estimate of about FAR = 30, but we are not very confident about this number being the correct number. The true FAR value could be 5, or it could be 50.

Quantification of model uncertainty is not normally covered by the risk analysis. But some examples exist where model uncertainty is assessed, see Section 2.1.3.

In practice it is difficult to perform a complete uncertainty analysis within this setting. In theory an uncertainty distribution on the total model and parameter space should be established, which is impossible to do. So in applications only a few marginal distributions on some selected parameters are normally specified, and therefore the uncertainty distributions on the output probabilities are just reflecting some aspects of the uncertainty. This makes it difficult to interpret the produced uncertainties.

Bayesian updating is a standard procedure for updating the uncertainty distribution when new information becomes available. See Appendix A.3 and Section 4.3.4 for a description of this procedure.

Figure 2.3 summarizes the main features of the classical approach with uncertainty quantification. It is also known as the probability of frequency framework, see Apostolakis and Wu (1993) and Kaplan (1992). In this framework the concept of probability is used for the subjective probability and the concept of frequency is used for the objective probability based on relative frequency. When the analyst assesses uncertainties related to **q**, he or she will often need to make simplifications, such as using independence.

Here are the main steps of this approach:

1. Identify suitable risk indices.
2. Develop a model of the activity or system being analysed, linking more detailed elements of the system and the overall risk indices.
3. Estimate unknown parameters of the model.
4. Establish uncertainty distributions for the parameters of the model.
5. Propagate them through the model to obtain uncertainty distributions for the risk indices.
6. Establish predictive distributions and estimates of the risk indices.

In the rest of this section we look at the use of sensitivity and importance analysis, and risk acceptance and tolerability. The starting point is a classical approach using best estimates or a classical approach including uncertainty analysis.

Sensitivity and importance analysis

It is common to combine the above approaches with sensitivity analyses. A sensitivity analysis is a study of how sensitive the risk is with respect to changes

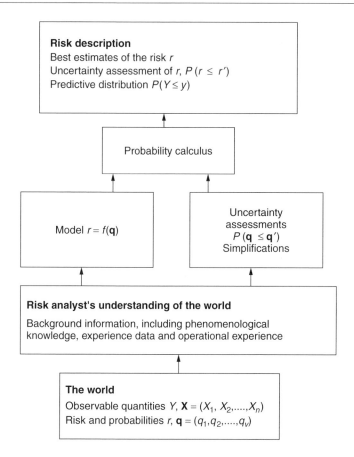

Figure 2.3 Basic elements of a risk analysis. Classical approach with uncertainty assessments

in input parameters of the risk model. Let us return to the offshore installation example. Then we can show how the FAR value estimate changes as a function of varying the leakage frequency λ. One factor is changed at a time. A λ value equal to 1 gives a FAR estimate of 32. If the λ value is reduced to 0.5, the estimate of FAR is reduced to 16, and if the λ value is increased to 2, the estimate of the FAR value becomes 64. We observe that the FAR estimate is proportional to the value of λ. In most cases the parameters are varied over a broad range; this is to identify the importance of the parameter and its improvement potential. Probability estimates may be set to their extremes, 0 and 1. It is common to use this way of thinking to rank the importance of the various elements of the system, for example safety barriers. An alternative approach that is also used for importance identification, is to look for the effect of small changes: How quickly does the risk index change when the input parameter changes? The measure is specified by taking the partial derivative of the risk index with respect to the parameter.

In this way we can derive two importance measures from a sensitivity analysis. In applications we often see that sensitivity analyses are mixed with uncertainty analyses. But a sensitivity analysis is not an uncertainty analysis as the analyst does not express his or her uncertainty related to the possible values of the parameters. A sensitivity analysis can be used as a basis for an uncertainty analysis. By presenting the result as a function of a parameter value, the analyst and the decision-makers can evaluate the result in view of uncertainty in the parameter value, but the sensitivity analysis *alone* does not provide any information about the uncertainties of the parameter value.

Risk acceptance and tolerability

Risk analysis is often used in combination with risk acceptance criteria, as inputs to risk evaluation. The criteria state what is deemed as an unacceptable level of risk. The need for risk-reducing measures is assessed with reference to these criteria. In some industries and countries it is a requirement in regulations that such criteria should be defined in advance of performing the analyses. Two main categories of quantitative risk acceptance criteria are in use:

Absolute values

- The probability p of a certain accidental event should not exceed a certain number p_0. Examples: the individual probability that a worker shall be killed in an accident during a specific year should be less than 10^{-3}; the probability of a safety function impairment during a specific year should not exceed 10^{-3}.
- The statistical expected number of fatalities per 100 million exposed hours, i.e. the FAR value, shall not exceed a certain number m_0.

Three regions

- The risk is so low that it is considered negligible.
- The risk is so large that it is intolerable.
- An intermediate region where the risk shall be reduced to a level which is as low as reasonably practicable (ALARP).

Consider absolute values. To avoid unnecessary repetitions, we will focus on evaluating the FAR value.

In this case the risk is considered acceptable if and only if the FAR value is less than or equal to m_0. In practice an estimate FAR* is used since the true value of FAR is unknown. Remember that the probabilistic framework is classical. The normal procedure is to use this estimate to decide on the acceptability of risk. Thus no considerations are given to the uncertainty of the estimate FAR*. Consider the offshore installation example again and suppose the risk acceptance criterion is equal to FAR = 50. The best estimate was FAR* = 32, meaning that risk-reducing measures are not required. But the true risk could be much higher than 50, as demonstrated by the uncertainty analysis on page 18. According to this analysis, the analysts have computed a subjective probability of 31% for the true FAR value to be higher than 50. So just ignoring the uncertainties, as is done when adopting the best-estimate approach, does provide an effective tool in that it produces clear recommendations but these recommendations could be rather poor, as demonstrated by this example. Nevertheless, this approach is often seen in practice. To cope with the uncertainty problem, standardized models and input data are sought. The acceptance criterion is considered to be a function of the models and the input data. This means that we have to calibrate the acceptance criteria with the models and the input data. The chosen model and the estimates of the model parameters are assumed to be equal to the true model and the true parameters. As long as we stick to these models and input data, we can focus on the best estimates and we need not be concerned about uncertainties. Apparently, this approach functions quite well as long as we are not facing novel problems and situations, e.g. due to new technology. Then it is difficult to apply this way of thinking. And, of course, the uncertainty problem is not solved; it is just ignored to produce an efficient procedure for expressing acceptable or unacceptable risk.

Risk acceptance criteria should therefore be used with care. They should be regarded more as guidelines than as requirements. A limit for what is acceptable risk related to human lives and environmental issues could prove there is a strong commitment from management, but it may sometimes reduce flexibility to achieve cost-effective arrangements and measures. When decisions that concern risk are to be made, costs and benefits will always be considered. What is acceptable risk has to be seen in relation to what we can achieve by accepting the risk.

This type of reasoning is more in line with the ideas of the three-regions approach. This approach is considered attractive by many since it allows consideration of costs and benefits. Chapter 5 illustrates how the cost-benefit considerations can be carried out. The three-regions approach is typically used in relation to a best-estimate approach. The above discussion on absolute values also applies here, as there are two defined limits against which to compare the risk. Sometimes the ALARP region is called an uncertainty region. But it is not clear how we should understand this uncertainty region. Here is one possible interpretation, where we assume that risk is expressed by the estimate FAR* of the true value of FAR. Simple numerical values are used to illustrate the ideas.

If FAR* is less than 1, we conclude that risk is negligible. If FAR* is larger than 100, we conclude that risk is intolerable, and risk-reducing measures are required. Now suppose we have indicated an uncertainty factor 10 for the estimate FAR*. Then if FAR* is larger than 100, we have strong evidence that the true value FAR is larger than $100/10 = 10$. Similarly, if the estimate FAR* is less than 1, we have strong evidence that the true value FAR is less than $1 \times 10 = 10$. Thus 10 represents the *real* criterion for intolerance and negligibility, respectively. The interval [1,100] is an uncertainty region where the ALARP principle applies. Decision-makers can draw conclusions about intolerability (above 100) or acceptance/negligibility (below 1), with the intermediate region interpreted as tolerable only if risk reduction is impracticable (which means cost-benefit considerations).

Although such an interpretation seems natural, we have not seen it often expressed in precise terms in applications.

2.1.3 Reliability Analysis

A reliability analysis can be viewed as a special type of risk analysis or as an analysis which provides input to the risk analysis. In this section we briefly review the standard approach for conducting reliability analysis. As this approach is similar to the one described in the previous section, we will just introduce the main features of reliability analysis and refer to Section 2.1.2 where appropriate. We distinguish between a traditional reliability analysis and methods of structural reliability analysis, as they represent different traditions, the former dominated by statisticians and the latter by civil engineers.

Traditional reliability analysis

To illustrate the ideas, we use a simple example. Figure 2.4 shows a so-called fault tree and its associated block diagram for a system comprising three components, where component 3 is in series with a parallel system comprising components 1 and 2. We may think of this system as a safety system of two components in parallel, meaning that both components (1 and 2) must be in a failure state to obtain system failure. Component 3 represents a common-mode failure, meaning that the occurrence of this event causes system failure. The AND and OR symbols represent logic gates. In an OR gate the output event occurs if one of the input events occurs. In an AND gate the output event occurs if all of the input events occur.

Each component is either functioning or not functioning, and the state of component i ($i = 1, 2, 3$) is expressed by a binary variable X_i:

$$X_i = \begin{cases} 1 & \text{if component } i \text{ is in the functioning state} \\ 0 & \text{if component } i \text{ is in the failure state.} \end{cases}$$

Similarly, the binary variable Φ indicates the state of the system:

$$\Phi = \begin{cases} 1 & \text{if the system is in the functioning state} \\ 0 & \text{if the system is in the failure state.} \end{cases}$$

We have in this case

$$\Phi = \Phi(\mathbf{X}) = [1 - (1 - X_1)(1 - X_2)]X_3, \tag{2.3}$$

where $\mathbf{X} = (X_1, X_2, X_3)$, i.e. the state of the system is determined completely by the states of the components. The function $\Phi(\mathbf{X})$ is called the *structure function* of the system, or simply the structure. From this three-component system it is straightforward to generalize to an n-component system.

Figure 2.4 is an example of a so-called monotone system, because its performance is not reduced by improving the performance of a component. More

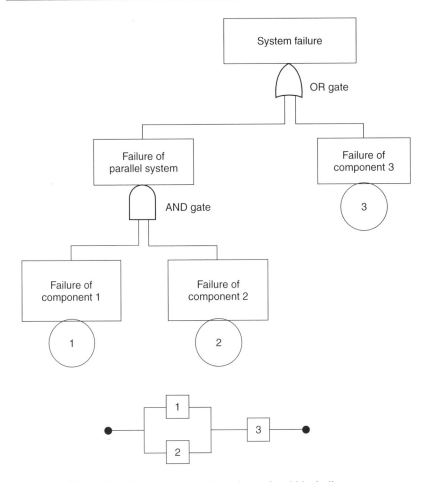

Figure 2.4 Fault tree example and associated block diagram

precisely, a monotone system is a system having a structure function Φ that is non-decreasing in each argument, and if all the components are in the failure state then the system is in the failure state, and if all the components are in the functioning state then the system is in the functioning state. All the systems we consider are monotone.

Let

$$p_i = P(X_i = 1), i = 1, 2, \ldots, n,$$
$$h = h(\mathbf{p}) = P(\Phi(\mathbf{X}) = 1), \qquad (2.4)$$

where $\mathbf{p} = (p_1, p_2, \ldots, p_n)$. It is assumed that all components are functioning or not functioning independently of each other. The probability p_i is called the reliability of component i. The system reliability h is a function of the component reliabilities $\mathbf{p}$, and this function is called the reliability function. Parametric lifetime models are often used to express p_i, for example an exponential model

$1 - e^{-\lambda_i t}$, where λ_i is the failure rate of the component and t is the time of interest. If T_i is a random variable having this distribution, we may think of T_i as the time to failure of this component. So component i functioning at time t is the same as having $T_i > t$, hence $p_i = e^{-\lambda_i t}$.

In a reliability analysis the system reliability h is calculated given the component reliabilities p_i. Let us look at the three-component example first. The reliability of the parallel system of components 1 and 2, h_p, is given by

$$h_p = 1 - P(X_1 = 0)P(X_2 = 0) = 1 - (1 - p_1)(1 - p_2),$$

noting that both components must be in the failure state to ensure that the system is in the failure state. This parallel system is in series with component 3, meaning that both the parallel system and component 3 must function for the system to function. It follows that the reliability of the system h is

$$h = [1 - (1 - p_1)(1 - p_2)]p_3.$$

This could also have been seen directly from (2.3) as

$$h = P(\Phi(\mathbf{X}) = 1) = E\Phi(\mathbf{X})$$
$$= E[1 - (1 - X_1)(1 - X_2)]X_3$$
$$= [1 - (1 - p_1)(1 - p_2)]p_3.$$

Now consider a practical case where a reliability analysis is to be conducted. The questions we ask are similar to those in Section 2.1.2:

- How is reliability expressed?
- What is the meaning of probability and reliability?
- How is uncertainty understood and addressed?
- What is the meaning of a model?
- How are parametric probability models like the exponential model understood and used?

The answers are analogous to those in Section 2.1.2. The situation is similar but with $h(\mathbf{p})$ in place of $f(\mathbf{q})$. A classical approach is most common. The best-estimate approach means providing best estimates $\widehat{\mathbf{p}}$ of $\mathbf{p}$ and using the model $h(\mathbf{p})$ to generate best estimates of the system reliability, i.e. $\widehat{h} = h(\widehat{\mathbf{p}})$. The classical approach with uncertainty analysis means that uncertainty distributions are generated for the parameters $\mathbf{p}$, and through the model $h(\mathbf{p})$ this uncertainty is propagated through the system to obtain an uncertainty distribution over the system reliability h. Note that as h is a linear function in each p_i, we have

$$Eh(\mathbf{p}) = h(E\mathbf{p}),$$

where the integration is over the uncertainty distribution of $\mathbf{p}$. We have assumed independent uncertainty distributions for the p_is. To avoid repetition, we omit the details.

The reliabilities, the probability distributions and associated parameters are usually estimated by classical statistical methods but Bayesian methods are also popular. Refer to Appendix A for a brief summary of these methods. See also Chapter 4.

Methods of structural reliability analysis

Methods of structural reliability analysis (SRA) are used to analyse system failures and compute associated probabilities. The performance of the system is described by a so-called limit state function g, which is a function of a set of quantities (random variables) $\mathbf{X} = (X_1, X_2, \ldots, X_n)$. The event $g(\mathbf{X}) < 0$ is interpreted as system failure, meaning that the probability of system failure, the unreliability, is given by the probability $p_g = P(g(\mathbf{X}) < 0)$. As an example, we can think of $g(\mathbf{X}) = X_1 - X_2$, where X_1 represents a strength variable of the system and X_2 represents a load variable. If the load variable exceeds the strength variable, system failure occurs. The difference $X_1 - X_2$ is called the safety margin.

Often a set of limit state functions is logically connected as unions and intersections, leading to probabilities such as

$$P([g_1(\mathbf{X}) < 0 \cup g_2(\mathbf{X}) < 0] \cap g_3(\mathbf{X}) < 0).$$

If $\mathbf{X}$ has distribution function F, we can write

$$p_g = \int_{\{\mathbf{x}: g(\mathbf{x}) < 0\}} dF(\mathbf{x}).$$

If F has a density f, this integral takes the form

$$\int_{\{\mathbf{x}: g(\mathbf{x}) < 0\}} f(\mathbf{x}) \, d\mathbf{x}.$$

Methods of SRA are used to calculate the probability p_g, and to study the sensitivity of the failure probability to variations of parameter values. Often Monte Carlo simulation is used, but this is sometimes a rather time-consuming technique. An alternative approach to finding p_g is to use an approximate analytical method, for example FORM or SORM. These methods give sufficiently accurate results in most cases. We refer to textbooks on SRA for further details; see also Section 4.4.3. It is common to assume that the distribution F has a parametric form, and often a multivariate normal distribution is used. Consider for example the load strength model mentioned earlier. Assuming that the pair (X_1, X_2) is a multivariate (bivariate) normal distribution with $EX_i = \mu_i$ and $\mathrm{Var} X_i = \sigma_i^2$, $i = 1, 2$, and a correlation coefficient ρ, it follows that the limit state function $X_1 - X_2$ also has a normal distribution; its mean is equal to $\mu_1 - \mu_2$ and its variance is equal to $\sigma_1^2 + \sigma_2^2 - 2\rho\sigma_1\sigma_2$.

As for the risk analysis community, the probabilistic basis for the analyses is seldom precisely specified. The underlying thinking is, however, along the lines of the classical approach, with best estimates, the use of confidence intervals, and uncertainty analyses of unknown parameters and calculation of the predictive distribution of the failure event. Returning to the load strength model, the classical approach with uncertainty analysis means that uncertainty distributions for the parameters μ_i, σ_i^2 and ρ are established. Let H denote the distribution for all five parameters. Then if $F(\mathbf{x}|\mu_1, \sigma_1^2, \mu_2, \sigma_2^2, \rho)$ denotes the distribution of $\mathbf{X}$ given the parameters, we can calculate the predictive distribution of the failure event as follows:

$$P(X_1 - X_2 < 0) = \int p_g(\mu_1, \sigma_1^2, \mu_2, \sigma_2^2, \rho)\,\mathrm{d}H(\mu_1, \mu_2, \sigma_1^2, \sigma_2^2, \rho), \quad (2.5)$$

where

$$p_g(\mu_1, \sigma_1^2, \mu_2, \sigma_2^2, \rho) = \int_{\{\mathbf{x}:g(\mathbf{x})<0\}} \mathrm{d}F(\mathbf{x}|\mu_1, \sigma_1^2, \mu_2, \sigma_2^2, \rho). \quad (2.6)$$

In this way F reflects the stochastic (aleatory) uncertainty and H the state-of-knowledge (epistemic) uncertainty.

Modelling uncertainty is an important issue in structural reliability analysis. Starting from the limit state function model g, the uncertainty related to g could be incorporated by introducing an error variable X such that the updated limit state function can be written as Xg. Seeing X as a state variable, we are back to the standard set-up presented above. We will return to this thinking in Section 4.4.3.

2.2 ECONOMIC RISK

2.2.1 General Definitions of Economic Risk in Business and Project Management

In economic applications a distinction has traditionally been made between certainty, risk and uncertainty, based on the availability of information. Certainty exists if the outcome of a performance measure is known in advance. Risk and uncertainty relate to situations in which the performance measures have more than one possible outcome, and the outcome is not known in advance. Under risk the probability distribution of the performance measures can be assigned objectively, whereas under uncertainty these probabilities must be assigned or estimated on a subjective basis.

Reference is often made to two risk situations, one in which probability is deduced objectively as in gambling situations where all the possible outcomes are assigned the same probability, and one in which probability is accurately estimated from relevant empirical data as in actuarial and insurance settings. For the uncertainty situation we can interpret the probabilities as measures of uncertainty, as subjective probabilities expressing degrees of belief. Alternatively, the probabilities can be interpreted as subjective estimates of true, underlying, objective probabilities. In most cases the level of precision in the literature allows both these interpretations.

In the earlier literature on risk a distinction was often made between speculative risk and pure risk. Speculative risk refers to situations where the outcomes of the performance measures of interest could be either favourable or unfavourable. Petroleum prices and the production potential of a petroleum reservoir are examples. The pure risk concept refers to situations where the outcomes of the performance measure are purely unfavourable. Examples include certain types of accident events causing loss of life, damage to the environment, or loss of assets or financial interests. However, in real life, damage to one party is often followed by growth among others. For example, an accident occurring in one company could create a more favourable market position for other companies, or accidents might create new business opportunities. Thus we cannot say that uncertainty related to the occurrence of accidents is solely associated with unfavourable outcomes, and the concept of pure risk cannot be used generally for typical undesirable events.

Within the area of project management, the term 'uncertainty' expresses lack of ability to predict the outcome of a performance measure. Probability is used to express the uncertainty related to what will be the true outcome. By this definition, probability is a subjective probability. This is, however, not so clear when reading the project management literature. When establishing probability distributions, reference is made to subjective probabilities, empirical distributions or theoretical distributions more or less 'verified' by use of empirical data.

A commonly used distribution for expressing uncertainty is the normal distribution $N(\mu, \sigma)$, which is specified by the parameters μ and σ, the mean and standard deviation of the distribution, respectively. Many analysts and researchers in the field talk about estimates of μ and σ, and they discuss the legitimacy of the assumptions related to the use of a particular distribution. Their probabilistic basis is not clear. Is their starting point a relative frequency interpretation of probability and their analysis a search for accurate estimates, or is the probability distribution simply a subjective probability expressing uncertainty about an unknown quantity?

The term 'risk' is often given the same definition as uncertainty – lack of ability to accurately predict the outcome of a performance measure. More narrow definitions are also applied, for example that risk is equal to the statistically expected value of the performance measure when only the possible negative outcomes are considered. To illustrate this definition, consider the case where the performance measure C can take four values, either $C = -5$, $C = -1$, $C = 1$, or $C = 2$, and the associated probabilities are 0.05, 0.20, 0.50, 0.25. Then, according to this definition,

$$\text{Risk} = -E[\min\{0, C\}] = 0.45.$$

The possible positive outcomes are reflected in the term 'opportunity', which is defined as the statistically expected value of the performance measure when only the possible positive outcomes are considered. For the above numerical example, we obtain

$$\text{Opportunity} = E[\max\{0, C\}] = 1.0.$$

The overall expected value of C, $E[C]$, is equal to 0.55.

Utilities and decision-making

In a decision-making context, risk is sometimes defined in relation to a utility function reflecting the worth of various possible losses or consequences (outcomes). Let X be a random variable representing the possible outcomes associated with a decision or act, and let $u(X)$ be the utility function. Then

$$\text{Risk} = -E[u(X)].$$

Now, starting from a 'rational' preference ordering on the outcomes, it can be shown that this leads to the use of expected utility as the decision criterion; see Savage (1962), von Neumann and Morgenstern (1944) and Bedford and Cooke (2001). In practice the expected utility theory of decision-making is used as follows. One assigns probabilities and a utility function on the set of outcomes, and then uses the expected utility to define the preferences between actions. These are the basic principles of what is known as rational decision-making. In this paradigm, utility is as important as probability. It is the ruling paradigm among economists and decision analysts.

The notion of utility is used to express the concept of risk aversion. We call the decision-maker's behaviour risk averse if $E[u(X)] < u(E[X])$. The behaviour is called risk neutral if $E[u(X)] = u(E[X])$. Risk aversion is the standard behavioural assumption, which implies that uncertainty is considered to be an unfavourable phenomenon.

We refer to the final section of this chapter and Chapter 5 for further details on the meaning and use of utilities, and a discussion of some of the above conventions, and in particular the rational decision-making paradigm.

2.2.2 A Cost Risk Analysis

A cost risk analysis is a tool typically used in project risk management. The purpose of the analysis is to estimate the project cost and provide an evaluation of the uncertainties involved. To this end, a model is developed linking the total cost of the project to a number of subcosts, expressing costs related to different work packages. As an illustration we will consider a simple case where the total cost C can written as the sum of two cost quantities C_1 and C_2, i.e.

$$C = C_1 + C_2. \tag{2.7}$$

A cost estimate of C is obtained by summing cost estimates of C_1 and C_2. This is straightforward. In addition to the cost estimate, an uncertainty interval is normally produced. Assuming normal distributions, a 68% uncertainty interval is established by the cost estimate $\pm$ the standard deviation σ_C; a 95% uncertainty interval by the cost estimate $\pm 2\sigma_C$. If the C_i are considered to be independent, this standard deviation is obtained from the standard deviation of C_i, denoted by σ_{C_i}, $i = 1, 2$, using the formula

$$\sigma_C = \sqrt{\sigma_{C_1}^2 + \sigma_{C_2}^2},$$

which is derived from the fact that the variance of a sum of independent random variables is equal to the sum of variances of the random variables. If dependency is to be incorporated, the standard deviation σ_C is adjusted so that

$$\sigma_C = \sqrt{\sigma_{C_1}^2 + \sigma_{C_2}^2 + 2\rho\sigma_{C_1}\sigma_{C_2}},$$

where ρ is the correlation coefficient between C_1 and C_2. Consider a case where the cost estimate is 5.0 for both C_1 and C_2, the standard deviations for C_1 and C_2 are 1.0 and 2.0 respectively, and the correlation coefficient is 0.5. Then a cost estimate of 5.0 ± 2.6 is reported, with confidence about 70%. The cost estimates and standard deviations are established using experience data whenever they exist. Expert judgements are also used.

Often these uncertainty intervals are referred to as confidence intervals, but they are better described as uncertainty intervals or prediction intervals because they provide statements about the observable costs, not the expected costs that form the basis of confidence intervals.

In many applications the uncertainty is specified as relative values of the costs. Suppose the C_i are judged to be independent and the cost estimates for C_1 and C_2 are 2 and 3, respectively. Furthermore, suppose that the uncertainty is $\pm 50\%$ relative to the costs, i.e. ± 1 and ± 1.5 respectively, with a confidence of $3\sigma_{C_i}$, which is about 0.999. Then the cost estimate of 5 is presented with a reduced uncertainty of $\pm 36\%$, as $3\sigma_C$ is given by

$$\sqrt{(3\sigma_{C_1})^2 + (3\sigma_{C_2})^2} = \sqrt{1^2 + 1.5^2} = 1.8,$$

which is 36% of 5.

In practice Monte Carlo simulation often is used. As mentioned in Section 2.1.2, this is a computer-based technique that is used to generate realizations of the system or activity (here the cost quantities) being analysed, and based on these realizations the desired probability distributions can be established. When using Monte Carlo simulation, distributions other than the normal distribution can easily be handled, such as triangle distributions, and complex dependency structures can be incorporated.

2.2.3 Finance and Portfolio Theory

The research and development in economic risk theory has put much attention on portfolio theory – the relationship between the risk related to a portfolio of a number of securities (e.g. stocks or projects) and the individual risk of the securities comprising that portfolio. This theory is closely linked to the capital asset pricing model (CAPM).

The future values of the securities are unknown quantities, or random variables. The mean value of the portfolio is simply the sum of the mean values of the individual securities in the portfolio. The variance, which is the most common measure of risk in this setting, is the sum of the variances of the individual securities plus a term reflecting the covariance between the values of the securities. To see this more precisely, consider the value V of a portfolio of N

securities, the ith having value X_i and weight $1/N$. Then the relative value of the portfolio, V, which can be written $(X_1+X_2+\cdots+X_N)/N$, has a variance of

$$\text{Var}[V] = \frac{1}{N}\overline{\text{Var}} + \left(1 - \frac{1}{N}\right)\overline{\text{Cov}}, \qquad (2.8)$$

where $\overline{\text{Var}}$ is average variance of the individual securities, i.e.

$$\overline{\text{Var}} = \frac{\text{Var}X_1 + \text{Var}X_2 + \cdots + \text{Var}X_N}{N},$$

and $\overline{\text{Cov}}$ is the average covariance between pairs of securities, i.e.

$$\overline{\text{Cov}} = \frac{2\sum_{i<j}\text{Cov}[X_i, X_j]}{N(N-1)}.$$

The first term on the right-hand side of (2.8) expresses the non-systematic risk and the second term the systematic risk. The non-systematic risk emerges from marginal uncertainty embodied in the values of the single securities, for example from the possible occurrence of accidental events. The investor can remove this uncertainty by diversification, i.e. investments in securities from a number of companies in various industries. Systematic risk is uncertainty in the value of a security, which cannot be removed by diversification. It is generated by general market forces, political events, etc., which affect a significant number of companies in the market.

Now, when N is large, the variance (2.8) of the portfolio is approximately equal to the average covariance. Thus the non-systematic risk is negligible when N is sufficiently large, and the portfolio risk is ruled by the systematic risk. Often the so-called β factor is used to express this risk. More precisely, the factor β_i is defined as the covariance between a market portfolio of value X_M and security i having value X_i, divided by the variance of the market portfolio, Var_M, i.e.

$$\beta_i = \frac{\text{Cov}[X_i, X_M]}{\text{Var}_M} = \frac{\rho_{iM}\sigma_i}{\sigma_M}, \qquad (2.9)$$

where ρ_{iM} is the correlation coefficient between the security i and the market, σ_i is the standard deviation of X_i, and σ_M is the standard deviation of X_M. The higher systematic the risk related to a security, the higher the expected return required by the investors. The main conclusion of CAPM is that the price of security i will adjust to reflect the risk, so its expected return is given by

$$E[r_i] = r_f + \beta_i(E[r_m] - r_f), \qquad (2.10)$$

where r_f is the risk-free discount rate and r_m is the return from the market as a whole. The quantity r_i is the sum of dividends received and capital gains. Suppose we have these figures:

Price of security at beginning of period = 100
Price of security at the end of period = 110
Dividends received during period = 5

Then we obtain a return r_i given by

$$r_i = \frac{(110 - 100) + 5}{100} = 0.15.$$

Equation (2.10) shows how CAPM determines $E[r_i]$ as the sum of the risk-free rate of return and β_i multiplied by the so-called risk premium of the market, $E[r_m] - r_f$. The β value can be interpreted as the number of systematic risk units. Thus the risk cost contribution is expressed by the risk premium of the market multiplied by the number of units of systematic risk.

In practice the risk measure β is determined based on historic stock prices and market indices. It can be computed as the slope of a regression line between periodic (usually yearly, quarterly or monthly) rates of returns for the market portfolio (as measured by a market index) and the periodic rates of return for security i as follows:

$$r_{ij}^* = \alpha_i^* + \beta_i^* r_{mj}^* + \varepsilon_i, \tag{2.11}$$

where ε_i is a random error term, r_{ij}^* is the periodic rate of return for security i, r_{mj}^* is the periodic rate of return for the market index, α_i^* is a constant term determined by the regression and β_i^* is the computed historical beta for security i given by

$$\beta_i^* = \frac{\sum_j (r_{mj}^* - \bar{r}_m^*)(r_{ij}^* - \bar{r}_i^*)}{\sum_j (r_{mj}^* - \bar{r}_m^*)^2};$$

here $\bar{r}_m^*$ and $\bar{r}_i^*$ are the means of r_{mj}^* and r_{ij}^*, respectively. The terms α_i^* and β_i^* are the values of α_i and β_i that minimize

$$\sum_j (r_{ij}^* - \alpha_i - \beta_i r_{mj}^*)^2.$$

Part of the basis for CAPM is the assumption of efficient markets, i.e. markets in which all relevant information is reflected in the price of the security. However, real stock markets are not completely efficient and other analysis tools, such as fundamental and technical analysis are used to obtain information relevant for the future development of the value of the stocks. Fundamental analysis focuses on the economic forces behind supply and demand that cause stock prices to increase, decrease or stay the same. Technical analysis studies market actions. Movements in the market are used to predict future changes in stock price. In short, fundamental analysis studies the cause of market movements and technical analysis studies the effect of market movements.

A diversified investor is only concerned with systematic risk. Thus accident risk as studied in Section 2.1 is of little concern from a portfolio risk viewpoint as most accidents will only affect a few companies, or perhaps just one, not the market as a whole. And the topic of economic accident risk has not been paid much attention in business risk contexts. Exceptions are the methods used to calculate insurance premiums in insurance companies.

2.2.4 Treatment of Risk in Project Discounted Cash Flow Analysis

In selection and management of projects, the net present value (NPV) is the most common performance measure. To measure the NPV of a project, the relevant project cash flows are specified, and the time value of money is taken into account by discounting future cash flows by the required rate of return. The formula used to calculate NPV is

$$\text{NPV} = \sum_{t=1}^{T} \frac{X_t}{(1+r)^t}, \tag{2.12}$$

where X_t is equal to the cash flow at year t, T is the time period considered (in years) and r is the required rate of return, or the discount rate. The terms 'capital cost' and 'alternative cost' are also used for r. As these terms imply, r represents the investor's cost related to not employing the capital in alternative investments. When considering projects where the cash flows are known in advance, the rate of return associated with other risk-free investments, like bank deposits, forms the basis for the discount rate to be used in the NPV calculations.

When the cash flows are uncertain, which is usually the case, various approaches are taken. They can be summarized as follows:

- Represent the cash flows X_t by their mean values $E[X_t]$ and increase the required rate of return to outweigh the possibilities for unfavourable outcomes.
- Express uncertainty related to the cash flows and apply the risk-free discount rate r_f.
- combine the first two items by expressing uncertainty in the cash flows and discounting with a risk-adjusted rate of return.

When dealing with uncertainty in project cash flows, we distinguish between systematic and non-systematic risk to the investor (see the previous section), who is commonly assumed to be a shareholder in possession of a well-diversified portfolio of securities. In projects, systematic risk (market risk or non-diversifiable risk) refers to uncertainty in factors affecting the cash flow that are also related to other activities in the market such as energy prices, raw material prices and political situations. Non-systematic risk is uncertainty in cash flow factors solely impacting the specific project, such as operational delays, accidental events, dependency on critical personnel, production rate of a specific oil well and the demand for a new niche product. It will not affect other investments made by the investor. Since the impact of non-systematic risk to the value of the investor's portfolio can be more or less eliminated by diversification, the systematic risk is the main focus in studies of project profitability, e.g. NPV analysis.

The first of the three approaches mentioned above is the standard procedure in NPV calculations of uncertain projects. It applies a risk-adjusted rate of return, usually determined on the basis of CAPM; see the previous section. Equation

(2.10) shows how CAPM determines the expected return from a security, $E[r_i]$, as the sum of the risk-free rate of return and β_i multiplied by the risk premium of the market, $E[r_m] - r_f$. The quantity β_i is a measure of the systematic risk associated with the activity of company i, and is determined by the covariance between the value of the company and the market relative to the total variance in the market; see equation (2.9). The β value is usually determined on basis of historical data from similar projects or from the industry sector to which the project belongs. Using this approach, the greater the systematic risk associated with the company's activities, the higher the discount rate. This corresponds to the principle of risk aversion: when uncertainty (systematic risk) is large, this must be compensated by a higher return requirement. High discount rates imply a greater reducing effect on the value of cash flows, the more distant they are in the future. This also takes into account that the cash flows which are the most distant are often the most uncertain (risky). That all investors are risk averse, is one of the assumptions underpinning CAPM.

The second approach, where the analysts express their uncertainty about the cash flows and discount with the risk-free rate of interest, exists in numerous variants. A common procedure is the scenario analysis, in which the cash flow of the project is usually estimated in three cases: the pessimistic, the most probable and the optimistic. Probabilities are assigned to reflect the uncertainty regarding which scenario will occur, and this forms the basis for weighing the NPVs derived in each case. Another widely used method, which requires more extensive description of uncertainties by probabilities, is Monte Carlo simulation. The profit of a project may depend on a vast number of different quantities, and in such a simulation the uncertainties related to these quantities can be taken into account.

The third approach uses the same methods as described for approach 2, but the risk-adjusted discount rate, usually CAPM based is applied.

In most cases, under the three approaches, r is represented by a single number. Some analysts, however, choose to express r by a probability distribution, in order to reflect that a range of numbers might apply, depending on the relative weighting of the various arguments involved in the assessment of a proper r.

For scenario analysis and simulations it is argued by many economists that the risk-free discount rate should be used (the second approach), as the risk aspects of the NPV are summarized in the generated distribution. The uncertainty should not be accounted for a second time, by using a risk-adjusted discount rate. The interpretation of the distribution of NPV is widely discussed in the literature; see Myers (1976) and also Hull (1980).

> If NPV is calculated using an appropriate risk adjusted discount rate, any further adjustment for risk is double-counting. If a risk-free rate of interest is used instead then one obtains a distribution of what the project's value would be tomorrow if all uncertainty about the project's cash flows were resolved between today and tomorrow. But since uncertainty is not resolved in this way the meaning of the distribution is unclear. Hull (1980), p. 33.

Others claim, however, that the risk-adjusted rate of return should be used (third approach), since the simulations in most cases only reflect some of the uncertainty involved, or since most of the probabilities reflect unsystematic risk, not covered by the β measure.

The choice of discount rate does not, however, seem to present major practical problems under all applications. Some companies merely focus on questions such as, What is the probability (uncertainty) for the project to provide more than $y\%$ return? If $y\%$ is used as the discount rate, the answer to this question is simply the probability of NPV being greater than zero.

A performance measure closely related to NPV is the internal rate of return (IRR), which is defined at the rate of return i such that r is equal to i if NPV $= 0$. In many respects, a distribution for IRR is more useful than a distribution for NPV in answering questions such as the one above.

Instead of presenting the whole probability distribution of the NPV or IRR, it is common to report only the mean, e.g. E[NPV], and one or more measures of spread like the variance Var[NPV], the standard deviation SDV[NPV], the coefficient of variation SDV[NPV]/E[NPV] or a specific quantile in the distribution. These measures of spread are referred to as risk measures.

In the above setting, probability is used to express the uncertainty related to what will be the true outcome, so probability is by definition a subjective probability. But, as stated above, this is not so clear when reading the project management literature. When establishing a probability distribution, reference is made to subjective probabilities, empirical distributions or theoretical distributions more or less 'verified' by use of empirical data.

2.3 DISCUSSION AND CONCLUSIONS

2.3.1 The Classical Approach

We are not very much attracted by the classical approach to risk and risk analysis as seen in engineering applications. The problem is the introduction of and focus on fictional probabilities. These probabilities exist only as mental constructions, they do not exist in the real world. An infinite population of similar units need to be defined to make the classical framework operational. This probability concept means that a new element of uncertainty is introduced, the true value of the probability, a value that does not exist in the real world. Thus we are led to two levels of uncertainty and probability, which in our view reduces the power of risk analysis. We are interested in the behaviour of the units under consideration. What the classical approach can give is just some uncertainty statements about averages over fictional populations. We feel that this approach has the wrong focus. It does not give a good basis for supporting decisions.

For the populations introduced, it is supposed that they comprise similar units. The meaning of the word 'similar' is rather intuitive, and in some cases it is obvious what is meant. In other cases, the meaning is not clear at all. Let us look at an example.

Consider the probability of at least one fatality during one year in a production facility. According to the classical relative frequency view, this probability is interpreted as the proportion of facilities with at least one fatality when considering an infinite number of similar facilities. This is of course a thought experiment – in real life we just have one such facility. Therefore, the probability is not a property of the unit itself, but the population it belongs to. How should we then understand the meaning of similar facilities? Does it mean the same type of buildings and equipment, the same operational procedures, the same type of personnel positions, the same type of training programmes, the same organizational philosophy, the same influence of exogenous factors, etc. As long as we speak about similarities on a macro level, the answer is yes. But something must be different, because otherwise we would get exactly the same output result for each facility, either the occurrence of at least one fatality or no such occurrence. There must be some variation on a micro level to produce the variation of the output result. So we should allow for variations in the equipment quality, human behaviour, etc. But the question is to what extent we should allow for such variation. For example, in human behaviour, do we specify the safety culture or the standard of the private lives of the personnel, or are these factors to be regarded as factors creating the variations from one facility to another, i.e. the stochastic (aleatory) uncertainty, using the terminology from Section 2.1? We see that we will have a hard time specifying what should be the framework conditions of the experiment and what should be stochastic uncertainty. In practice we seldom see such a specification carried out, because the framework conditions of the experiment are tacitly understood. As seen from the above example, it is not obvious how to make a proper definition of the population.

We recognize that the concept 'similar' is intuitively appealing, although it can be hard to define precisely. But the main problem with the classical approach is not this concept, it is the fact that risk is a constructed quantity that puts focus in the wrong place, on measuring fictional quantities.

2.3.2 The Bayesian Paradigm

Bayesian methods are often presented as an alternative to the classical approach. But what is the Bayesian alternative in a risk analysis context? In practice and in the literature we often see a mixture of classical and Bayesian analyses, see Section 2.1.2. The starting point is classical in the sense that it is assumed there exists an underlying true risk. This risk is unknown, and subjective probability distributions are used to express uncertainty related to where the true value lies. Starting by specifying probability distributions on the model parameter level, procedures are developed to propagate these distributions through the model to the risk of the system. Updating schemes for incorporating new information are presented using Bayes' formula. We have referred to this basis as the classical approach with uncertainty analysis. As mentioned in Section 2.1.2, this approach is also called the probability of frequency framework, in which the concept of probability is used for the subjective probability and the concept of frequency is used for the objective relative frequency based probability.

This approach to risk analysis introduces two levels of uncertainty: the value of the observable quantities such as the number of failures of a system, the downtime, etc., and the correct value of the risk. The result is often that both the analysis and the results of the analysis are considered uncertain. This does not provide a good basis for communication and decision-making.

Now, how does this way of thinking relate to the Bayesian approach as presented in the literature, for example Barlow (1998), Bernardo and Smith (1994), Lindley (2000), Singpurwalla (1988) and Singpurwalla and Wilson (1999)? As we see from these references and others, and from Chapter 4 and Appendix A, the Bayesian thinking is in fact not that different from the probability of frequency approach described above. The point is that the Bayesian approach, as presented in the literature, allows for fictional parameters, based on thought experiments. These parameters are introduced and the uncertainty in them is assessed. Thus, from a practical viewpoint, an analyst would probably not see much difference between the Bayesian approach as presented in the literature and the probability of frequency approach referred to above. Of course, Bayesians would not speak about true, objective risks and probabilities, and the predictive form is seen as the most important one. However, in practice, Bayesian parametric analysis is often seen as an end-product of a statistical analysis. The application and understanding of probability models focuses on limiting values of quantities constructed through a thought experiment, which are very close to the mental constructions of probability and risk used in the classical relative frequency approach.

In our view, applying the standard Bayesian procedures, gives too much focus on fictional parameters, established through thought experiments. The focus should be on observable quantities. We believe there is a need for a rethinking of how to present the Bayesian way of thinking, to obtain a successful implementation in a practical setting. In a risk analysis comprising a large number of observable quantities, a pragmatic view of the Bayesian approach is required, in order to conduct the analysis. Direct probability assignments should be seen as a useful supplement to establishing probability models where we need to specify uncertainty distributions of parameters. A Bayesian updating procedure may be used for incorporating new information, but its applicability is in many cases rather limited. In most real-life cases we would not perform a formal Bayesian updating to incorporate new observations – rethinking of the whole information basis and approach to modeling is required when we conduct the analysis at a particular point in time, for example in the pre-study or concept specification phases of a project. Furthermore, we should make a sharp distinction between probability and utility. In our view it is unfortunate that these two concepts are seen as inseparable as is often done in the Bayesian literature.

The word 'subjective', or related terms such as 'personalistic', are well-established terms in the literature. However, as noted in the preface, we find such terms somewhat difficult to use in practice. We prefer to speak about probability as a measure of uncertainty, and make it clear who is the assessor of the uncertainty.

2.3.3 Economic Risk and Rational Decision-Making

As noted in Section 2.2.1, in economic risk theory, references are often made to literature restricting the risk concept to situations where the probabilities related to future outcomes are known, and uncertainty to the more common situations of unknown probabilities. This convention is in our view a blind alley and should not be used – it violates the intuitive interpretation of risk which is closely related to situations of unpredictability and uncertainty. In a framework based on subjective probabilities, known probabilities do not exist – all probabilities are subjective assessments of uncertainty – different assessors could produce different probabilities.

Economic risk is closely related to the use of utilities and rational decision-making. The optimization of the expected utility is the ruling paradigm among economists and decision analysts. We do recognize the importance of this paradigm – it is a useful decision-making tool in many cases. But it is just a tool, a normative theory saying how to make decisions strictly within a mathematical framework – it does not replace management review and judgement. There are factors and issues which go beyond the framework of utilities and rational decision-making, that management needs to consider. In practice there will always be constraints and limitations restricting the direct application of the expected utility thinking. Yet the theory is important as it provides a reference for discussing what good decisions are. The fact that people often violate the basis (axioms) of the theory – they do not behave consistently and coherently – is not an argument against this theory. The expected utility theory says how people ought to make decisions, not how they are made today. We may learn from the descriptive theory telling us how people actually behave, but this theory cannot replace normative theory. We do need some reference, even if it is to some extent theoretical, for the development of and for measurement of the goodness of decisions. In our view the expected utility theory can be seen as such a reference.

Cost-benefit analysis is another method for balancing costs and benefits. It is often used to guide decision-making in the ALARP region. The idea of the method is to assign monetary values to a list of burdens and benefits, and summarize the 'goodness' of an alternative by the expected net present value. The method is subject to strong criticism. The main problem is related to the transformation of non-economic consequences, such as (expected) loss of life and damage to the environment, to monetary values. What is the value of a (statistical) life? What is the value of future generations? These are difficult issues and have received much attention in the literature. There are no simple answers. The result is often that the cost-benefit analyses just focus on certain consequences and ignore others. Nevertheless, we find that this type of analysis provides useful insight and decision support in many applications. We are, however, sceptical about a *mechanical* transformation of consequences to monetary values, for in many cases it is more informative to put attention on each consequence separately and leave the weighting to management and the decision-maker, through a more informal review and judgment process. See Section 5.1.

As for risk analysis, the probabilistic basis for cost-benefit analysis is seldom clarified, but the classical thinking with a search for correct probability values seems to be dominant. It is common to question the validity of cost-benefit analyses because of their unrealistic assumptions about the availability of the data needed to complete the analyses. The underlying philosophy seems to be that without objective, hard data the analyses break down.

How does cost-benefit analysis relate to expected utility theory? Could we justify using one method in one case and the other method in a different case? These questions are important, but it is difficult to answer them using the standard decision theory. Either the utility theory is considered as the only meaningful tool, or this theory is rejected – it does not work in practice – and cost-benefit analyses are used.

2.3.4 Other Perspectives and Applications

The principles, methods and models presented in this chapter are related to engineering and business. But they are also used in other areas such as information and communication, biotechnology, agriculture, criminal law, food industry, medicine and occupational health. Within each area we find special nomenclature, conventions and procedures, but the same fundamental issues are being discussed:

- How do we express risk and uncertainty?
- How do we understand probabilities?
- How do we understand and use models?
- How do we understand and use parametric distribution classes and parameters?
- How do we use historical data and expert opinions?

These issues have been discussed in this chapter from different perspectives. Repeating the discussion for other application areas would be tedious and is omitted. To extend the range of applications, we have included some examples in Chapters 4 and 5 from areas outside engineering and business.

The classical approach to risk and risk analysis is dominating in many areas, such as medicine and occupational health. This is perhaps not so surprising as it is often possible in these areas to define large populations of 'similar units', for examples human beings. And then the traditional statistical approach seems to fit well. We can use the statistical techniques to 'prove' that a new medicine, for example, is superior to the old. There is a drive for proofs, scientific proofs, such that strong conclusions can be made. The lack of a humble attitude to knowing the truth about risk is, in our view, not only a problem in the engineering community; we also see it in for example medicine. We do understand social scientists and others that are provoked by the somewhat arrogant attitude among many analysts and scientists telling the world in situations involving large uncertainties that they know the truth and the non-experts are biased – they proclaim that the non-experts do not have the proper information and knowledge, and they are strongly influenced by perceptional factors such as dread. Yes, many people are strongly influenced by perceptional factors and lack proper information and knowledge about relevant topics. But they are not necessarily biased. There are uncertainties, meaning that there would be more than one possible direction. The issue is who we trust, who we listen to. We would weigh different judgements and views differently depending on the bases for holding them. Judgements having strong data and methodological support, plus consensus about the critical assumptions, would be given more weight than a layperson expressing his view without any reference to empirical evidence or theoretical reasoning. Our way of thinking has a scientific basis as far as we give reference to coherent and consistent judgements, but the processes of assessing uncertainties and decision-making have to be recognized as lying outside the classical natural science paradigm of controlled experiments.

When considering people's evaluations of, and behaviour towards, hazards, the term 'risk perception research' is often used as the generic label for this field of social science. Not only does it involve psychologists, it also takes input from a range of other disciplines, including sociology, anthropology, decision theory and policy studies. In this research, different definitions of risk are being used. We review two of the most common. The first, called the 'economic perspective', views risk in terms of a judgement about uncertainty. This might be an objective statistical probability but in most cases it is a subjective probability expressing degree of belief, or an evaluation of uncertainties covering aspects such as vagueness and ambiguity. Historically, in psychology there has been a long tradition of work that adopts this economic perspective to risk, where uncertainty can be represented as an objective probability. Here researchers (often known as behavioural decision researchers) have sought to identify and describe how people make decisions in the face of statistical and other types of uncertainty, together with the ways in which actual behaviour departs (or does not depart) from the formal predictions of normative economic theories such as the expected utility theory.

The second way of defining risk in the social sciences is broader. Here risk refers to the full range of beliefs and feelings that people have about the nature of hazardous events, their qualitative characteristics and benefits, and most crucially their acceptability. This definition is considered useful if lay conceptions of risk are to be adequately described and investigated. The motivation is that there are a wide range of multidimensional characteristics of hazards, rather than just an abstract expression of uncertainty and loss, which people appear to evaluate in performing perceptions, such that the risks are seen as fundamentally and conceptually distinct. Furthermore, these evaluations may vary with the social or cultural group to which a person belongs and the historical context in which a particular hazard arises, and they may also reflect aspects of the physical and human or organizational factors contributing to hazard, such as trustworthiness of existing or proposed risk management.

We do see the problem of having a narrow definition of risk, for example linked to the probability concept. Risk is obviously more than probabilities. On the other hand, a wide definition like the second one is considered problematic as it does not distinguish between our judgement about how the world would be in the future and how we value this future and our judgments about it. Both aspects are covered by this broad risk concept. We find it useful to separate them. Our solution, as presented in the coming chapters, is to distinguish between a broad qualitative definition of risk and more narrow quantitative definitions of risk measures. The qualitative definition, which basically says that risk is uncertainties about the performance of the system being studied (the world), makes it meaningful to talk about description, analysis, evaluation, perception and acceptance of risk, and these terms would together include the whole range of aspects listed above for the broad social science definition of risk.

Risk perception research has generated an impressive body of empirical data showing that human judgements of hazards and their benefits involve multiple qualitative dimensions related in quite subtle and complex ways. We briefly touch on aspects of this research. This book focuses in Sections 4.1.2 and 5.2.2 more on how we should approach risk and uncertainty, whereas risk perception research focuses more on describing how people actually think and behave. We have learned a lot from this research, ideas that provide a basis for the direction we recommend. Here are some of the important general lessons:

- Risk acceptance cannot be based on evaluations of expected values only. A more comprehensive risk picture is required.
- People are poor assessors of uncertainties if the reference is an objective, true statistical probability.
- Probability assignments (uncertainty assessments) are influenced by a number of factors.
- Perception, acceptance and tolerability of risk are influenced by a number of factors, such as dread and knowledge.
- There are significant individual and group differences in risk perception and acceptance.
- Risk perception and acceptance may be fundamentally related to social judgements of things such as responsibility, blame and trust in risk management and managers.

The risk perception research is concentrated on laypersons' perceptions. This book discusses how decision-makers and analysts (experts) should approach risk and uncertainty, and laypersons' risk perception and acceptance is just one of the many factors to be considered when making decisions, see Chapter 5. The risk analyst's assessment of uncertainty using subjective probabilities is discussed in Chapter 4.

2.3.5 Conclusions

The alternative to the classical approach to risk analysis is the Bayesian approach, where the concept of probability is used as the analyst's measure of uncertainty or degree of belief. But this alternative approach has not been commonly accepted; there is still a lot of scepticism among many risk analysts when speaking about subjective probabilities. Perhaps one reason for this is lack of practical implementation guidelines. When studying the Bayesian paradigm, it is not clear how we should implement the theory in practice. We find the Bayesian literature very technical and theoretical. The literature is to a large extent concerned about mathematical and statistical aspects of the Bayesian paradigm. The more practical challenges of adopting the Bayesian approach are seldomly addressed.

We see the need for a rethinking of how to present the Bayesian approach to uncertainty and risk in a practical setting. The aim of the coming chapters is to present a basis for this thinking and to give guidelines and recommendations for an alternative presentation that addresses the criticisms we have raised.

BIBLIOGRAPHIC NOTES

Most textbooks on reliability and risk analysis are in line with the classical way of thinking as described in this chapter. They focus on estimation of reliability and risk, and if uncertainty is addressed, it means expressing confidence interval or subjective uncertainty distributions for relative frequency interpreted probabilities or expectations. Examples of books in this category are Henley and Kumamoto (1981), Høyland and Rausand (1994) and Vose (2000). Most of these books focus on methods of analysis and management. Foundational issues are not a main topic. Most applied risk and reliability analysts have been trained in such methods, but they have not spent very long reflecting on the foundations, even though many papers address this topic. Examples of such papers are Apostolakis (1990), Apostolakis and Wu (1993), Kaplan (1991, 1992), Kaplan and Burmaster (1999), Singpurwalla (1988, 2002), Aven and Pörn (1998) and Aven (2000a, 2000b). Several special issues of risk journals have been devoted to foundation, and in particular, aspects of uncertainty. They include special issues of the journal *Reliability Engineering and System Safety*; see Apostolakis (1988) and Helton and Burmaster (1996). G. Apostolakis and S. Kaplan have done pioneering work in establishing and discussing an appropriate basis for risk analysis. Probability of frequency thinking was introduced more than 20 years ago (Kaplan and Garrick 1981). Our presentation of the different categories of classical thinking is based on more recent work, e.g. Aven and Pörn (1998) and Aven and Rettedal (1998); it represents a rethinking of some of the basic ideas of Kaplan and others. In his work, Apostolakis compared the probability of frequency ideas and the more modern version of the Bayesian approach (Apostolakis and Wu 1993), and he pointed to the problem of introducing true but unknown frequencies. Our work in this area has been greatly inspired by the work of Apostolakis.

For an overview of the literature on sensitivity analysis, see Tarantola and Saltelli (2003) and Sandøy and Aven (2003).

The discussion on risk and tolerability is taken from Aven and Pitblado (1998). The economic risk review is partly based on Aven *et al.* (2003). Some basic references addressing economic risk, and in particular finance and portfolio theory, are Levy and Sarnat (1990) and Moyer *et al.* (1995). Project risk is addressed by Kayaloff (1988) and Nevitt (1989) among others. The common qualitative definition of risk in this context is lack of ability to predict the outcome of a performance measure. The more narrow definition of risk – the expected value of the performance measure when restricting attention to negative outcomes – is also popular; see Levy and Sarnat (1972), Levy (1998) and Jordanger (1998). The traditional definition of risk and uncertainty in Section 2.2.1 is mentioned by a number of textbooks, e.g. Douglas (1983).

Our presentation of the traditional reliability analysis is based on Barlow and Proschan (1975) and Aven and Jensen (1999). Methods for structural reliability analysis are reviewed by Ang and Tang (1984), Madsen et al. (1986), Toft-Christensen and Baker (1982) and Melchers (1987). Our presentation of the SRA methods is partly based on Aven and Rettedal (1998).

A basic reference for cost risk analysis is Vose (2000). Statistical decision theory and the use of utility theory are thoroughly discussed in Chapter 5 and the relevant literature is listed there.

Many social scientists have criticized traditional engineering risk assessments. We mention Beck (1992), Douglas and Wildavsky (1982), Perrow (1984) and Shrader-Frechette (1991). The critical point seems to be that the idea of an objective risk cannot be justified. According to Slovic (1998), risk does not exist out there, independent of our minds and cultures. We must take the 'naive positivist' view, to use the terminology of Shrader-Frechette (1991), that risk exists objectively and can be measured, and replace it by a more balanced view. The answer is not the other extreme – the relativist view saying that A's risk description is as good as B's, regardless their bases – but a middle position, expressing that formal risk assessments provide useful information to support decision-making, by combining facts and judgements using scientific principles and methods. Most people, we think, are in favour of such a middle position, see (Shrader-Frechette 1991), but the challenge is to establish a proper platform for it. The aim of this book is partly to provide one.

There is an enormous literature on risk perception research. We refer to Okrent and Pidgeon (1998), Pidgeon and Beattie (1998) and the references therein. Our review of risk perception research is based on Pidgeon and Beattie (1998).

The foundational literature on subjective probabilities links probability and decisions; see Ramsey (1926) and de Finetti (1972, 1974). By observing the bets people make or would make, one can derive their personal beliefs about the outcome of the event under consideration, see Section 5.1.2. This view of subjective probabilities was disputed by Koopman (1940); see also Good (1950), who holds a more 'intuitionist' view on subjective probabilities. The intuitive thesis says that probability derives directly from intuition and is prior to objective experience. Intuitionists consider that the Ramsey–de Finetti 'revealed belief' approach is too dogmatic in its empiricism as, in effect, it implies that a belief is not a belief unless it is expressed in choice behaviour. We agree with the intuitionists on this point, and make a sharp distinction between probability assignments and decision-making. This distinction seems also to be common among many applied Bayesian risk analysts. Our view of probability is explained in detail in the coming chapters.

According to the Bayesian paradigm, there are no true objective probabilities. However, a consistent subjectivist would act in certain respects as if such probabilities do exist. The result is that many analysts just as easily assume that the true objective probabilities exist as well as the subjective ones, see Good (1983: 154). In our terminology, they shift from the Bayesian paradigm to the probability of frequency approach.

3

How to Think about Risk and Risk Analysis

This chapter presents a unifying approach to risk and risk analysis based on the idea that risk is a way of expressing uncertainty related to future observable quantities. Section 3.1 gives the main ideas. Sections 3.2 and 3.3 give examples to illustrate these ideas.

3.1 BASIC IDEAS AND PRINCIPLES

This section presents the basic principles of the unifying approach to risk and risk analysis. The starting point is an activity or a system that we would like to analyse now to provide decision support for investment, design, operation, etc. The interesting quantities in the future are the performance of the activity or system (from now on referred to as the system), measured by profit, production, production loss, number of fatalities, the occurrence of an accident, and so on. These are the quantities that we should like to know the value of at the time of the decisions since they provide information about the performance of the alternatives. Unfortunately, though, these quantities are unknown at the time of the decision-making. Thus we are led to predictions of these quantities, reflecting in some sense, what are to be expected. But these predictions would normally not provide sufficient information; assessment of uncertainties is required. We need to see beyond expectations. The expected value could give a prediction of 1.5, but the actual outcome of the quantity could for example be 0, 5, 100. Assessments of uncertainties related to each possible outcome would give additional and useful information compared to just reporting the expected value. To express our uncertainties, we need a measure and probability is our answer. The reference is a certain standard such as drawing a ball from an urn. If the possible outcomes are 0, 5 and 100, we may assign probability figures, say 0.89, 0.10 and 0.01, respectively, corresponding to the degree of belief we have in the different values. We may also use odds; if the probability of an event A is 0.10, the odds against A are 9:1. The assignments are based on available information and knowledge; if we had sufficient information, we would be able to predict with certainty the value of the quantities of interest. The quantities are unknown to us as we have lack of knowledge about how people would act, how machines would work, etc. Systems analysis and modelling would increase the knowledge and thus hopefully reduce uncertainties. In some cases, however, the analysis and modelling could in fact increase our uncertainty about the future value of the unknown quantities. Think of a situation where the analyst is confident that a certain type of machine is to be used for future operation. A more detailed analysis may, however, reveal that also other types of machine are being considered. And as a consequence, the analysts's uncertainty about the future performance of the system may increase. Normally we would be far away from being able to see the future with certainty, but the principle is the important issue here – uncertainties related to the future observable quantities are epistemic, that is, they result from lack of knowledge.

These are the main principles of the unifying approach. They are summarized in the following list and illustrated in Figure 3.1.

Basic principles

1. Focus is placed on quantities expressing states of the 'world', i.e. quantities of the physical reality or the nature, that are unknown at the time of the analysis but will, if the system being analysed is actually implemented, take some value in the future, and possibly become known. We refer to these quantities as *observable* quantities.
2. The observable quantities are predicted.
3. Uncertainty related to what values the observable quantities will take is expressed by means of probabilities. This uncertainty is *epistemic*, i.e. a result of lack of knowledge.
4. Models in a risk analysis context are deterministic functions linking observable quantities on different levels of detail. The models are simplified representations of the world.

Figure 3.1 is read as follows. A risk analyst (or a risk analyst team) conducts a risk analysis. Focus is on the future performance of the system (the world), and in particular some observable quantities reflecting the performance of the system, Y and $\mathbf{X} = (X_1, X_2, \ldots, X_n)$. Based on the analyst's understanding of the world, the analyst develop a model (several models) that relates the overall system performance measure Y to $\mathbf{X}$, which is a vector of quantities, on a more detailed level. The analyst assesses uncertainties of $\mathbf{X}$, and that could mean the need for simplifications in the assessments, for example using independence between the quantities X_i. Using probability calculus, the uncertainty assessments of $\mathbf{X}$, together with the model g, give the results of the analysis, i.e. the assigned probability distribution of Y, and a prediction of Y. The uncertainty distribution of Y and $\mathbf{X}$ are known as predictive distributions.

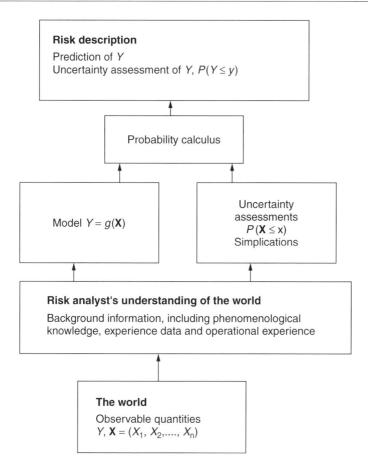

Figure 3.1 Basic elements of a risk analysis

The above principles express the main features of our thinking. This thinking is primarily motivated by a pragmatic concern: how to make the analysis functioning in practice, a search for structure and simplicity, and ease of communication. We recommend assigning probabilities only for observable quantities.

The typical steps of a risk analysis following these principles can be summarized as follows:

1. Identify the overall system performance measures (observable quantities on a high level).
2. Develop a deterministic model of the system linking the system performance measures and observable quantities on a more detailed level.
3. Collect and systemize information about these low-level observable quantities.
4. Use probabilities to assess uncertainty of these observable quantities.
5. Calculate the uncertainty distributions of the performance measures and determine suitable predictions from these distributions.

Sometimes a model is not developed as the analysis is just a transformation from historical data to an uncertainty distribution and predictions related to a performance measure; steps 2 and 4 can then be ignored. Often the predictions are derived directly from the historical data without using the uncertainty distributions. Although we have the main focus on the high-level performance measures, the uncertainty assessments of the low-level observable quantities are also of interest, as they provide valuable insights about key elements of the system.

In this approach, risk is qualitatively defined as uncertainty related to the performance of the analysis object, the system. In other words, risk is uncertainty about the world. In the quantitative analysis, uncertainty is expressed by probabilities related to the observable quantities $Y, X_1, X_2, \ldots$. Risk is associated with the whole distribution of the observable quantities (performance measures). Summarizing measures such as the mean, the variance and quantiles are risk measures which can give more or less information about risk.

In the following we discuss in more detail some of the key elements of the approach; see Figure 3.1. We normally use Y and $Y_i, i = 1, 2, \ldots$, to express observable quantities on a high system level and $X_i, i = 1, 2, \ldots$, to express observable quantities on a more detailed system level. When not using this nomenclature, it will be clear from the context what are high-level observable quantities and what are low-level observable quantities.

3.1.1 Background Information

All probabilities are conditioned on the background information (and knowledge) that we have at the time we quantify our uncertainty. This information covers historical system performance data, system performance characteristics (policies, goals and strategies of a company, types of equipment to be used, etc.), knowledge about the phenomena in question (fire and explosions, human behaviour, etc.), decisions made, as well as models used to describe the world. Assumption is an important part of this information and knowledge. We may assume for example in an accident risk analysis that no major changes in the safety regulations will take place for the time period considered, the plant will be built as planned, the capacity of an emergency preparedness system will be so and so, an equipment of a certain type will be used, etc. These assumptions can be viewed as frame conditions of the analysis and the produced probabilities must always be seen in relation to these conditions. If one or more assumptions are dropped, this would introduce new elements of uncertainty to be reflected in the probabilities. Note, however, that this does not mean the probabilities are uncertain. What are uncertain are the observable quantities. For example, if we have established an uncertainty distribution $p(c|d)$ over the investment cost c for a project, given a certain oil price d, it is not meaningful to talk about uncertainty of $p(c|d)$ even though d is uncertain. A specific d gives one specific probability assignment, a procedure for determining the desired probability. By opening up for uncertainty assessments in the oil prize d, more uncertainty is reflected in our uncertainty distribution for c, using the law of total probability.

The point is that in our framework uncertainty is only related to observable quantities, not assigned probabilities. See Section 4.2 for further comments on this issue, in the context of model uncertainty.

For the sake of simplicity we normally omit the dependency on the background information when writing probabilities. This should not create any confusion as long as the background information is not varying throughout the discussion.

3.1.2 Models and Simplifications in Probability Considerations

In the above predictive approach, cost models linking the cost of various cost elements and the total cost, and models like event trees, fault trees and limit state functions, are developed to improve predictions of Y; uncertainty is assessed on a detailed level using relevant information, and this gives the uncertainty distributions and predictions related to Y. So a model in this setting means a deterministic model. See Sections 3.2 and 3.3 and Section 4.2 for further discussion on the use of models in this setting. To conduct a risk analysis it is often necessary to make some simplifications of the uncertainty assessments, i.e. the probability considerations, for example by using independence for a number of random quantities.

3.1.3 Observable Quantities

The quantities focused on are observable, meaning that they express states of the world. The value of an observable quantity is well defined as conventions and procedures exist expressing how to measure it. No ambiguity can be present. Thus an observable quantity has a true, objective value. For example, the number of fatalities in a company during a specified period of time would clearly be an observable quantity. If we consider the number of injuries, it is not so obvious. We need to define precisely what an injury means. And according to such a definition, we would have one correct value. The fact that there could be measurement problems in this case – some injuries are not reported – does not change this. The point is that a true number exists according to the definition and if sufficient resources were made available, that number could be found. This example illustrates that observable quantities include cases where we could better describe the quantities as potential observable quantities. Here is another example that makes this point clear. A production company produces units, say mobile telephones, and suppose we focus on the proportion of units that fail during a certain period of time and according to a certain definition of failure, among all produced units in one year for one particular type of mobile telephone. This proportion is potentially observable, since it can be measured exactly if sufficient resources are made available. In practice that would not normally be done. Yet we classify it as observable.

Now, what about a relative frequency? Is such a quantity observable? Well, the answer is both no and yes. Consider as an example a case where the system is a production facility and we focus on the occurrence of an accidental event (suitably defined) for a one-year period. Then we can define a relative frequency probability by the proportion of similar production facilities where this event occurs. If this population of similar production facilities is just a thought experiment, it is fictional, then this relative frequency is not observable. We will not be able to observe the relative frequency in the future – it is not a state of the world. If, however, such a population can be specified, the relative frequency can be viewed as observable. Such a population is difficult to imagine in this case unless we extend the meaning of similar to include every type of production facility. Then we would be able to obtain a value of the proportion of facilities where this event occurs, but that proportion would not be very relevant for the system we study. What is a real population and what is a fictional population need to be determined in each application. As a general rule we would say that populations may exist when we deal with repeatable games, controlled experiments, mass-produced units and large physical populations like human beings, etc. For the mobile telephone example above, a population can be defined and the relative frequency, i.e. the proportion of failed units, is an observable quantity. However, this book concentrates on other types of application, where the system is unique in the sense that we cannot find reasonably similar systems without doing thought constructions.

Let p denote an observable relative frequency. We refer to it as a chance. It is an objective property of the sequence or population considered – it is not a probability for the assessor, though were p known to the assessor, it would be the assessor's probability for any event in the sequence or in the population. Note that there is a fundamental distinction between uncertainty that involves judgement by the assessor and is described by probabilities, and uncertainties, or better, variations, that are properties of the world external to the assessor. See Chapter 4, p. 79 for some further comments on the link between chances and our predictive approach.

As a final remark related to a quantity being observable, consider the volume produced in some units for a gas production system during a certain period of time, say one year. For all practical purposes, stating that this volume is for example 2.5 would be sufficiently accurate. If you go into the details, the exact production volume could be somewhat difficult to define and measure, but thinking practically, and using the conventions made for this kind of measurement, the correctness of the measurement, for example 2.5, is not an issue. If it were, then more precise measurements would have been implemented.

3.2 ECONOMIC RISK

3.2.1 A Simple Cost Risk Example

A risk analyst in a company is to assess the investment cost Y for a development project related to a production facility. First he would like to make an assessment based on historical records from 20 rather similar development projects, which show a mean cost of 100 and an empirical standard deviation of 30. Note that these numbers are not estimates of any underlying parameters in a probabilistic model – they are just summarizing measures found adequate for describing the data. Now, how should he approach the problem? Well, according to the predictive approach of Section 3.1, he should make a prediction Y and assess uncertainties. The mean of the historical data, 100, would be the natural candidate for the prediction of the project cost Y. He would present this number as a prediction of Y stressing that this prediction is based on the figures seen for the 20 other facilities. To express uncertainties the analyst may use a histogram distribution as shown in Figure 3.2.

A parametric distribution class may also be used to express uncertainties, such as the normal distribution or the lognormal distribution. Suppose in this case we would like to use the normal distribution (below we comment on the use of a lognormal distribution). To determine the distribution we need to specify the mean μ and standard deviation σ. The natural candidates would be the empirical quantities, i.e. 100 and 30, respectively. Thus a 95% prediction interval is given by $\mu \pm 2\sigma$, i.e. [40,160]. Thus the analyst has assigned a 95% probability for the future investment cost to be in the interval [40,160], based on the historical data. There is no uncertainty related to this interval. There is no meaning in speaking of uncertainty of the parameters μ and σ because such reasoning would presuppose the existence of true probabilities which do not exist in this setting. Consequently, the phrase 'estimation of these parameters' should be avoided as it indicates that we aim at coming as close as possible to true, underlying parameter values. What is uncertain is the future investment cost Y, and it is meaningless to discuss the correctness of the use of a normal distribution.

Suppose the company management is particularly interested in the event that the cost Y exceeds 160. Based on the above analysis, the analyst would predict

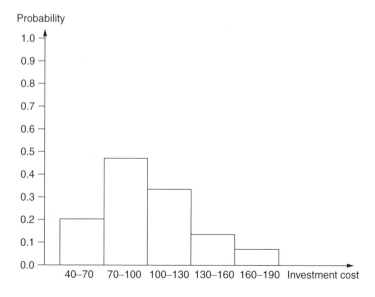

Figure 3.2 Uncertainty distribution for the investment cost

no occurrence of this event, and express the uncertainty by a probability of 2.5%.

To reflect the shape of Figure 3.2, it would have been more natural to use the lognormal distribution in place of the normal distribution. However, the procedure would have been analogous to the normal distribution case, replacing the observed cost values with natural logarithms and then computing the empirical mean and variance.

Mathematically, when using a parametric distribution class, this procedure is identical to producing estimates of parameters in a classical statistical context. But the way of thinking is different. We may produce the same normal distribution, but the meaning of that distribution is not the same. Uncertainty in our setting is related to the value of Y, whereas the classical approach would need to address uncertainty of the estimators relative to underlying true values. If we use the uncertainty distribution expressed by Figure 3.2, even a classically oriented statistician would probably find it confusing and disturbing to discuss uncertainty of this distribution relative to the underlying true distribution. But as soon as a parametric distribution class is introduced, the question about accuracy of the estimates is addressed. A parametric distribution in this context is just a mathematical class of functions that we consider suitable for expressing our uncertainty about observable quantities. There is no difference in principle between a histogram as shown in Figure 3.2 and the normal distribution with fixed values of μ and σ.

Of course, if our starting point had been an infinite (or very large) population of production facilities similar to the one analysed, we could speak about a true distribution of investment costs, as this distribution is observable, and the accuracy of the normal distribution as a model of this distribution. In the above case with historical records of 20 projects, such a population is not introduced as it means the introduction of a fictional population. If we were able to define an infinite or very large population of similar projects, we would have to extend the meaning of 'similar' to an extremely wide range of projects; the result is that the population becomes rather irrelevant for the facility studied.

The above approach to assessing the investment cost is based on rather limited information and knowledge, only the data for the 20 other facilities are taken into account. Thus large uncertainties are present. One way of reducing uncertainties and obtaining narrower prediction intervals, is to identify key factors related to the production facilities that are important for determining the cost. Suppose that the production volume is found to be a good indicator for the investment cost. The analyst then plots the investment cost as a function of the production volumes for the 20 facilities. To make this simple, suppose that the data fit well to a straight line and let $y = a + bx$ represent this line, where the constants a and b have been determined for example by least square regression (Appendix A.2.4), x is the input (independent) variable representing the production volume and y is the output (response) variable representing the investment cost. Now, based on a planned production volume x_0, we can use this line to obtain a prediction of the investment cost equal to $a + bx_0$. To express uncertainties in the investment cost given x_0, we may for example use a histogram like Figure 3.2 or a normal distribution. In the latter case, we may use a standard regression analysis to produce the empirical variance of the investment cost:

$$\frac{1}{n-2}[\sum(y-a-bx)^2,$$

where $n = 20$ and the sum is over the 20 observations of x and y.

These two approaches are very crude – the uncertainties and the prediction intervals are large. There is rather limited information on what the main contributors to the uncertainty are, what the effect of alternative arrangements are, etc. If such information is required and more confidence in the predictions is to be achieved, an analysis of the system needs to be conducted. Let us see how this can be done following the main steps listed in Section 3.1.

The performance measure in this case is the investment cost Y, which is defined according to standard economic conventions. Then a model is developed linking this investment cost and more detailed cost elements $X_i, i = 1, 2, \ldots, k$. In this case the model is simply the sum of the cost elements, that is

$$Y = \sum_{i=1}^{k} X_i.$$

As a predictor for the cost Y we would normally use the mean, EY, which is equal to the sum of the means (the predictors) of the various cost elements, as $EY = \sum_{i=1}^{k} EX_i$. This is straightforward; the challenge is to establish the uncertainty related to the value of Y. The uncertainty distribution can be established in different ways, as demonstrated in Section 2.2.2. The basic thinking can be summarized as follows. For each cost element C_i a probability distribution F_i is determined that expresses the analyst's uncertainty related to the value of X_i. This distribution is established based on historical data, if available, and the use of expert judgements. If a triangular distribution is used, we need to specify its minimum, its peak and its maximum. Then if the cost elements X_i are judged independent, the uncertainty distribution of Y is generated by the convolutions of the distributions of X_i. In practice the distribution of Y is often found by Monte Carlo simulation, in which values of X_i are generated according to its probability distribution. Often normal distributions are used to reflect uncertainties. Then the means and the variances need to be specified. If the cost elements are judged dependent, we also need to specify correlation coefficients. See Section 4.4.1 for further discussion of this example.

3.2.2 Production Risk

An oil company evaluates several design options for a gas production system. As a basis for the decision to be taken, it is of interest to obtain information about certain performance measures, for example the number of times the production is below demand in a certain period of time and the future production loss due to equipment failures and maintenance. Let Y_1 and Y_2 denote this number and this loss, respectively, for a given design alternative for the relevant period of time.

The loss could be expressed in millions of cubic metres of gas or normalized as a percentage in relation to the demanded volume. In the planning phase, Y_1 and Y_2 are unknown and we have to predict Y_1 and Y_2. The prediction can be done in different ways. We could compare with similar systems if available, or we could develop a more detailed model of the system reflecting the various subsystems and equipment; we develop a 'reliability model' of the system. Having established the model, uncertainties are restricted to the times to failure and the downtime durations of subsystems and equipment. Regardless of the approach taken, we will arrive at predictions of Y_1 and Y_2. The uncertainty related to the value of Y_1 and Y_2 we express through probabilities.

To see the basic elements of this framework in more detail, here are the details of such a reliability model. Assume that the system is a binary system of binary components, so that Y_1 is equal to the number of times the system fails and Y_2 is equal to the downtime of the system. This simplification is made to avoid too many technicalities. First we consider the case with one component only.

Let X_t represent the state of the component at time t; $X_t = 1$ if component i is functioning at time t and $X_t = 0$ if the component is not functioning at time t. We assume that the component is functioning at time 0, i.e. $X_0 = 1$. Let T_m, $m = 1, 2, \ldots$, represent the positive length of the mth operation period of the component, and let R_m, $m = 1, 2, \ldots$, represent the positive length of the mth repair time for the component, see Figure 3.3.

The following performance measures are defined:

$$Y_{1t} = \text{the number of failures in } [0, t]$$

$$Y_{2t} = \text{the downtime in } [0, t].$$

These quantities are both functions of the lifetimes and repair times. If S_k° denotes the time of completion of the kth repair, i.e. $S_k^\circ = \sum_{m=1}^{k}(T_m + R_m)$, with $S_0^\circ = 0$, we see that $Y_{1t} = k$ if $S_k^\circ - R_k \leq t$ and $S_k^\circ + T_{k+1} > t$. Furthermore, $Y_{2t} = \int_0^t (1 - X_s)\,ds$, where $X_s = 1$ if $S_k^\circ \leq s$ and $S_k^\circ + T_{k+1} > s$, and $X_s = 0$ otherwise.

This is modelling and it gives insight into the performance and the uncertainties. The remaining uncertainty is related to the values of the component lifetimes and repair times. The quantities T_m and R_m are unknown and we use

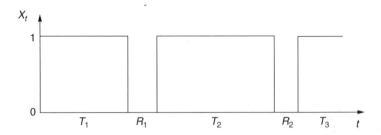

Figure 3.3 Time evolution of a failure and repair process for the component starting at time $t = 0$ in the operating state

probability distributions to express our uncertainty about what will be the true values. We judge all quantities T_m, R_m, $m = 1, 2, \ldots$, to be independent. This is a rather strong simplification, as we ignore learning by observing some of the lifetimes and repair times. But in some cases the background information is so strong that we could justify the use of independence; see Section 4.4.2.

We use the same distribution F for all uptimes and the same distribution G for all downtimes of the component. The finite means of these distributions are

$$\mu_F = ET_m \qquad \mu_G = ER_m.$$

The process X_t is a so-called alternating renewal process.

Now fix time t. Using the above models and the uncertainty distributions for the lifetimes and downtimes, associated uncertainty distributions for Y_{1t} and Y_{2t} can be computed, see Aven and Jensen (1999). Here we restrict attention to an example as an illustration. Suppose that $F(t) = 1 - e^{-\lambda t}$, where $\lambda = 1/19$ is the failure rate, and the repair time is a constant equal to 1. Further assume that $t = 100$. Then the computation is not so difficult. Let Y_{1t}^* be the Poisson process with rate $1/19$ generated by the uptimes of the component. Then we see that $P(Y_{1t} > k) \approx P(Y_{1t}^* > k)$ and $P(Y_{2t} > k) \approx P(Y_{1t}^* > k)$ ignoring the difference between calendar time and operational time. Exact formulas for $P(Y_{1t} > k)$ and $P(Y_{2t} > k)$ are given by

$$P(Y_{1t} > k) = P(S_{k-1}^\circ + T_k < t) = P(T_1 + \cdots + T_k < t - (k-1))$$
$$= P(Y_{1,t-(k-1)}^* \geq k),$$
$$P(Y_{2t} > k) = P(S_k^\circ < t) = P(T_1 + \cdots + T_k + k < t) = P(Y_{1,t-k}^* \geq k).$$

In the general case it is difficult to compute the uncertainty distributions for Y_{1t} and Y_{2t} and approximation formulas need to be used, see Aven and Jensen (1999). It is also common to use Monte Carlo simulation. When performing Monte Carlo simulations of Y_t, either Y_{1t} or Y_{2t}, we generate a sequence of independent, identically distributed random variables, say $Y_t^{(1)}, Y_t^{(2)}, \ldots, Y_t^{(k)}$, based on the same uncertainty distributions on the component level and the model linking Y_t and the component uptimes and downtimes. The simulation is performed in a classical statistical setting where the starting point is a probability that we wish to determine and we have repeated experiments generating independent and identically distributed random variables $Y_t^{(1)}, Y_t^{(2)}, \ldots, Y_t^{(k)}$. From this sample we arrive at the uncertainty distribution of Y_t, the mean and the variance of this distribution.

3.2.3 Business and Project Management

The standard approach to risk and risk analysis in business and project management is closely linked to the one described in Section 3.1. Consider the following simple example.

If you are going to invest a certain amount of money in a certain business, you are concerned about what the cash flow will be. Let Y denote this cash flow for a given period of time. Based on an evaluation of the cash flows for this business in previous years, you could make a prediction of Y. This is of interest to you, but you would also like to see an assessment of the uncertainties related to Y. This can be done in different ways, for example by expressing the probability of having a cash flow of at least y. This is a probability expressing uncertainty. Instead of considering the cash flow as such, we could investigate the number of time periods where the cash flow has been of a certain amount. This description could be useful to increase the information basis and make it easier to produce good predictions.

This presentation is in line with the principles of our unifying approach. But let us go one step further; suppose we express our uncertainty related to the value of Y by a normal distribution with mean μ and variance σ^2. According to our predictive approach, there is no meaning in speaking of uncertainty of these parameters unless they can be defined as observable. The situation is similar to the one discussed in Section 3.2.2. Consequently, the phrase 'estimation of these parameters' should be used with care as it indicates that we aim at coming as close as possible to true, underlying parameter values. What is uncertain is Y and it is meaningless to discuss the correctness of the use of a normal distribution if it is a subjective probability distribution and cannot be given a physical interpretation.

3.2.4 Investing Money in a Stock Market

Person s would like to invest 2 million dollars in a stock market. He considers two alternatives:

1. He buys stocks of type 1 only.
2. He buys stocks of type 1 and type 2, with 50% on each.

Before he decides what to do, he conducts a risk analysis according to the principles of Section 3.1. His focus is on the value of the stocks next year. Let us denote the next-year value of stocks of type 1 by X_1 and the corresponding next-year value for stocks of type 2 by X_2; both have a value of 1 million dollars today. Let Y_i denote the total value of the stocks next year for alternative i, $i = 1, 2$. Thus we have $Y_1 = 2X_1$ and $Y_2 = X_1 + X_2$. Person s looks at the historical records for the stocks, he analyses the corresponding companies' policies, strategies and plans for the future, and based on this information he predicts the future values of the stocks and assesses uncertainties. Normal distributions are used to express uncertainties. Hence it is sufficient to specify the means and the variances. Suppose his assessments give the same means; $EX_1 = EX_2 = 1.1$ (million), he predicts the same value for the two alternatives. This means the total value is the same; $2EX_1 = EX_1 + EX_2 = 2.2$. Furthermore, suppose that the variances are the same; $\text{Var}X_1 = \text{Var}X_2 = 0.04$. From this we see that

$$\text{Var}Y_1 = 4\text{Var}X_1 = 0.16$$

$$\text{Var}Y_2 = \text{Var}X_1 + \text{Var}X_2 + 2\text{Cov}(X_1, X_2) = 0.08 + 2\text{Cov}(X_1, X_2)$$

$$= 0.08 + \rho 0.08 = 0.08(1 + \rho),$$

where ρ is the correlation coefficient between X_1 and X_2. We conclude that the variance of Y_2 is smaller than the variance of Y_1, the difference depending on ρ. Thus if person s assigns a correlation coefficient ρ that is zero, the variance of Y_2 is just the half the variance of Y_1. Uncertainties (risk) are consequently smaller for alternative 2 than for alternative 1. But this does not lead to a recommendation on which alternative to choose. The risk presentation, here reported through the variance, is just an input to the decision-making. What is the best alternative must be seen in relation to policies, preferences and attitudes towards risk and uncertainty. Chapter 5 discusses the use of decision analyses to guide the decision-maker in situations like this. The important point is that uncertainty can be reduced by diversification, i.e. investments in stocks (securities) from a number of companies in various industries. This was discussed in Section 2.2.3.

Note that the analysis in this example does not depend on the use of a normal distribution, as long as we agree on using the variance as a way of representing the spread of the distribution.

If the analysis in this example had given different means and variances, the decision situation would have been more complex. Alternative 1 may have the highest mean and also the highest variance. To decide, we would need to take into account relevant policies, preferences and attitudes towards risk; see Chapter 5.

For an investor holding a diversified portfolio of securities, the mean of the uncertainty distribution related to the value of the securities is normally specified as the return from the securities in the market as a whole. The uncertainty distribution is then characterized by the spread of the distribution, and this spread can be measured by the variance and certain quantiles, for example.

3.2.5 *Discounted Cash Flow Analysis*

Refer to the cash flow analysis of Section 2.2.4. Under the approach to risk and risk analysis presented in Section 3.1, we see risk as uncertainty associated with observable quantities, and it is expressed in terms of probabilities related to these quantities. In cash flow analysis the cash flow components X_t are observable quantities, and probability distributions can be used to express associated uncertainties. Such distributions give a full description of risk related to a cash flow, according to our predictive approach. For a given discount rate r, the performance measure NPV is also an observable quantity, so the profitability of a project may be expressed by a probability distribution over the NPV, based on the distributions over the cash flows X_t. Thus risk in our setting means expressing probability distributions, or alternatively summarizing measures such as the mean and the standard deviation, over the NPV values for appropriate values of r.

We make a sharp distinction between risk and risk measures on the one hand, and decision rules based on risk and risk measures on the other hand. If we fix the discount rate r, as a risk-adjusted rate, compute the expected NPV and select the project alternative having the highest NPV value, we have introduced a decision rule based on a risk measure.

All the three procedures for NPV analysis discussed in Section 2.2.4 can be included in our predictive approach. We are in favour of expressing risk as uncertainty distributions over NPV for different values of r, to get insights, but we see the need for a simple rule to guide the decision-making based on the expected NPV value as explained above.

3.3 ACCIDENT RISK

Consider the offshore installation risk analysis example studied in Section 2.1.2, p. 11. We will look at how this study is conducted when we adopt the principles of Section 3.1.

The first task is to identify the overall system performance measures. From a personnel safety point of view, the objectives would of course be to avoid accidents, injuries and fatalities. From this we could formulate performance measures related to the occurrence of an accidental event, the number of injuries and the number of fatalities. Furthermore, performance measures related to the ability of the safety barriers to prevent escalation and reduce the consequences of a hazardous situation would be informative performance measures.

To simplify the analysis we focus on the possible occurrence of fatalities. Next we develop a deterministic model of the system, which in this example is just an event tree as shown in Figure 2.1. The tree models the possible occurrence of gas leakages in the compression module during a period of time, say one year. A gas leakage is referred to as an initiating event. The number of gas leakages is denoted by X. If an initiating event I occurs, it leads to Y fatalities, where $Y = 2$ if the events A and B occur, $Y = 1$ if the events A and not B occur, and $Y = 0$ if the event A does not occur. We may think of the event A as representing ignition of the gas and B as explosion. The model is very simple and is just used as an illustration of the ideas and principles of our predictive approach.

The model comprises some unknown, observable quantities which need to be studied. Let us first look at the number of leakages. Based on a review of relevant experience data we predict 4 leakages during one year. Uncertainties are reflected by a Poisson distribution with mean 4. This choice of uncertainty distribution is discussed in Chapter 4, p. 81. Given a leakage, only in rare cases the gas would ignite. Most leakages are small. Again modelling may be required. Such modelling would address the same type of aspect as mentioned in Section 2.1.2, p. 13, but the modelling approach would be different. The models developed could be explicitly formulated as deterministic functions, or they could be indirectly expressed by a procedure specifying our probability $P(A)$. A simple procedure would be to express $P(A|X = x)$, where X is the initial flow rate in kg/s, by the log-linear form

$$\log(P(A|X = x)) = a \log(x) + b,$$

for suitable x values, where a and b are constants, see Vinnem (1999: 130). Then by determining an uncertainty distribution for X, we arrive at our probability $P(A)$. More complex modelling would require development of models taking into account release characteristics, dispersion and ignition sources. Suppose that we arrive at a probability $P(A) = 0.002$, either using modelling or a direct argument using experience data and knowledge about the phenomena in question.

Similarly, we determine a probability $P(B|A)$. Let us suppose that we arrive at $P(B|A) = 0.2$. Then we can calculate the uncertainty distributions for the number of fatalities Y. We use approximation formulas like this:

$$P(Y = 2) = EX \cdot P(A) \cdot P(B|A). \qquad (3.1)$$

We can use this approximation because the event of two or more ignited leakages in one year has a negligible probability compared to the event of one ignited leakage. We obtain $P(Y = 2) = 0.0016$ and $P(Y = 1) = 0.0064$ and a FAR value equal to

$$[0.0016 \times 2 + 0.0064 \times 1] / [2 \times 8760] \times 10^8 = 55,$$

assuming 8760 hours exposure per year. The FAR value is defined as the expected number of fatalities per 100 million exposed hours.

The effect of proposed risk-reducing measures is in this case evaluated by assessing the effect on the probabilities. Suppose for example that some improved operational and maintenance procedures are implemented. A study of the possible causes of leakages might then result in an updated prediction of 2 leakages for one year, which would reduce the calculated risk by a factor of 2.

The analysis group concludes that the risk as calculated is rather high. Considering a ten-year period, a probability of an accident leading to fatalities is computed to be about 8%. Comparing this figure and the FAR value with risk numbers for similar activities, both risk analysis results and historical numbers, it is no doubt that the risk level as computed in this case is too high. Risk-reducing measures should therefore be considered.

Several leakages per year are to be expected. But given a leakage we would predict no loss of life. Most of the leakages represent no threat as they are very small. But if a large leakage should occur, the situation would be much more serious and fatalities could be the result. From this line of reasoning we see that by making event tree models for sizes of leakage, we could obtain a better understanding of what will happen given a leakage. Therefore this kind of division into categories is normally performed in practice.

The analysis group also needs to address possible increase in risk as a result of moving control out of the process area. The point is that, for certain types of scenarios, the operators would be able to detect deviations and implement corrective measures. We will not go further into this here.

Reliability analysis

Reliability analysis was introduced in Section 2.1.3. To see how the principles of Section 3.1 apply to these analyses, have a look at Section 4.4.3.

BIBLIOGRAPHIC NOTES

The presentation in this chapter is largely based on Apeland *et al.* (2001), Aven (2000a, 2000b, 2001), Aven *et al.* (2003). Our way of presenting how to approach risk and uncertainty is known as a predictive, Bayesian approach to risk and risk analysis, or as a predictive, epistemic approach.

This way of thinking, emphasizing observable quantities and using the risk analysis as a tool for prediction, is in line with the modern, predictive Bayesian theory as described in Bernardo and Smith (1994), Barlow (1998), Barlow and Clarotti (1993) and Spizzichino (2001). Our approach rewrites some established Bayesian procedures to obtain a successful implementation in a practical setting. Here are the essential points; they are further discussed in the next two chapters.

1. A sharp distinction is made between modelling to obtain better insights and predictions and the use of probability distribution classes to express uncertainty.
2. Fictional parameters are not introduced.
3. A rethinking of the whole information basis and approach to modelling is seen as an alternative to Bayesian updating.
4. A sharp distinction is made between probability and utility.

The importance of focusing on observable quantities has also been emphasized by others, such as Bedford and Cooke (2001), Morgan and Henrion (1990), Barlow and Clarotti (1993) and Geisser (1993).

Our definition of probability is in line with the one used by Lindley (1985, 2000); probability is a subjective measure of uncertainty, and the reference is a standard urn model. When referring to an observable relative frequency, we use the term 'chance'. A chance is closely linked to the concept of propensity, which is used to describe an objective probability representing the disposition or tendency of nature to yield a particular event on a single trial, see Popper (1959). Thus a propensity is a characterization of the experimental arrangement specified by nature, and this arrangement gives rise to certain frequencies when the experiment is repeated.

Keynes (1921) and other logical relationists insisted that there was less 'subjectivity' in epistemic probabilities than was commonly assumed. Keynes' point was that there is, in a sense, an 'objective' (albeit not necessarily measurable) relation between knowledge and the probabilities that are deduced from it. For Keynes, knowledge is disembodied and not personal. We disagree with this view on probability. Knowledge may be 'objective' in some sense, but probabilities cannot be separated from the person – probability reflects personal beliefs, it is subjective.

We refer to Bernardo and Smith (1994) and Lad (1996) for other key references on subjective probabilities and Bayesian theory.

4

How to Assess Uncertainties and Specify Probabilities

This chapter considers how to assess uncertainties and specify probabilities. Chapter 3 gave a number of examples demonstrating the way uncertainty can be assessed; now we go one step further and look in more detail into the assessment process. The key issue is how we arrive at a particular probability or probability distribution, using historical data, expert opinions and modelling. In particular, we study the case of uncertainty assessments of several quantities, for example some lifetimes of a component type or some cost elements in a development project. If we want to assess the uncertainty of say two lifetimes of units having the same type, how do we take into account the information gained on one lifetime by observing the other? The lifetimes should not be considered independent, but specifying a multivariate distribution is difficult. The question becomes when and how we can simplify the uncertainty assessments. Is independence appropriate in some cases, nonetheless?

Let Y denote the unknown, future observable quantity when only one quantity is of interest. We may think of the number of initiating events in an event tree, a cost element, or an indicator function that is equal to 1 or 0 depending on the outcome of a branching event of an event tree or an input event of a fault tree. The problem is to specify a probability distribution expressing our (the assessor's) uncertainty concerning the value of this quantity. This problem is the topic of Section 4.3. In Section 4.4 we address the multivariate case, i.e. specifying the distribution of the observable quantities $X_1, X_2, \ldots, X_n$, representing for example the lifetimes of n units or n cost elements. In Section 4.2 we discuss modelling, i.e. establishing a function g such that we can write $Y = g(X_1, X_2, \ldots, X_n)$ for some observable quantities $X_1, X_2, \ldots, X_n$. First, in Section 4.1 we will discuss what is a good probability assignment. Hopefully this discussion can provide some help when searching for guidance on which approach to use for specifying the distribution of Y.

Foundations of Risk Analysis T. Aven
© 2003 John Wiley & Sons, Ltd ISBN: 0-471-49548-4

4.1 WHAT IS A GOOD PROBABILITY ASSIGNMENT?

A probability in our context is a measure of uncertainty related to an observable quantity Y, as seen from the assessor's viewpoint, based on his state of knowledge. There exists no true probability. In principle an observable quantity can be measured, thus probability assignments can to some extent be compared to observations. We write 'in principle' as there could be practical difficulties in performing such measurements, see Section 3.1.3. Of course, one observation as a basis for comparison with the assigned probability is not very informative in general, but in some cases it is also possible to incorporate other relevant observations and thus give a stronger basis. Empirical control does not, however, apply to the probability at the time of the assignment. When conducting a risk analysis we cannot 'verify' an assigned probability, as it expresses the analyst's uncertainty prior to observation. What can be done is a review of the background information used as the rationale for the assignment, but in most cases it would not be possible to explicitly document all the transformation steps from this background information to the assigned probability. We conclude that a traditional scientific methodology based on empirical control cannot and should not be applied for evaluating such probabilities. We will elaborate on this later on.

It is impossible in general to obtain repeated independent measurements of assigned probabilities from the same individual because he is likely to remember his previous thoughts and responses. Consequently, there are no procedures for the measurement of the probability assignments that permit the application of the law of large numbers to reduce 'measurement errors'.

The difficulties involved in applying standard measurement criteria of reliability and validity to the measurement of probability assignment give rise to the question of how to evaluate and improve such assignments. Three types of criteria have been suggested: pragmatic, semantic (calibration) and syntactic.

4.1.1 Criteria for Evaluating Probabilities

The syntactic criterion is related to the probabilities obeying syntactic rules – the relations between assignments should be governed by the laws of probability. For example, if A and B are disjoint events, then the assigned probability of the event, A or B, should be equal to the sum of the assigned probabilities for A and B. A set of probability assignments is (internally) coherent only if it is compatible with the probability axioms. Coherence is clearly essential if we are to treat assignments of probabilities and manipulate them according to the probabilistic laws.

The pragmatic criterion is based on comparison with 'objective' values, the reality, and is applicable whenever the assigned probability of an event, e.g. the royal flush in poker, or a disease, can be meaningfully compared to a value that is computed in accordance with the probability calculus or derived from empirical data. For example, if history shows that out of a population of a million people, about two suffer from a certain disease, we can compare our probability to the rate $2/10^6$. However, such tests cannot be applied in most cases of interest as objective probabilities cannot be specified and sufficient relevant data are not available. Thus the pragmatic criterion is only rarely relevant.

Calibration tests relate to the ability to obtain correct statements when considering a number of assignments. Formally, a person is said to be well-calibrated if the proportion of correct statements, among those that were assigned the same probability, matches the stated probability, i.e. his hit rate matches his confidence. Clearly there is no way of validating, for example, a risk analyst's single judgement that the probability of a system to fail during a one-year period of operation is 0.1. But if the analyst is assessing many systems with a failure probability of 0.1, we would expect system failure to occur about 10% of the time. If, say, 50% of the systems fail, the analyst is badly calibrated. Often a scoring rule is used to reward a probability assessor on the basis of later observed outcomes. A simple scoring rule is the quadratic rule. If you assign a probability p for an event A, this rule gives the score $(1-p)^2$ if the event is true and p^2 if it is false.

Again, it is difficult to apply the criterion. The problem is that it does not apply at the point of assessment. The probability assignments are supposed to provide decision support, so the goodness of the probabilities needs to be evaluated before the observations. And the probabilities are assigned for alternative contemplated cases, meaning that comparisons with observations would be possible for just some of the probabilities assigned. There could also be changes in the background conditions of the probabilities from the assignment point to the observations. In risk analysis applications it often takes a long time before observations are available. The probabilities are in many cases small (rare events), which means that it is difficult to establish meaningful hit rates. Suppose we categorize probabilities in groups of magnitude 0.1 and 0.01 only. And suppose that we observe that for the two categories the risk analyst obtains 1 success out of 20 cases, and 0 out of 50 cases, respectively. Is the risk analyst then calibrated? Or to what extent is he calibrated? The hit rate for the first situation is 0.05, just a factor of 2 below the analyst's confidence; in the second situation, the hit rate is 0, which makes it difficult to compare with the probability assignments.

We conclude that calibration in general is not very useful to evaluate the goodness of the probability assignments in a risk analysis. Rather we see calibration as a tool for training risk analysts and experts providing input to the risk analysis, in probability assignments. By considering situations of relevance to the problems being analysed and where observations are available, we can evaluate the performance of the analysts and the experts, and improve their calibration in general. This training would increase the credibility of the risk analyst and the experts providing input to the risk analysis.

In situations where a number of probabilities are assigned and observational feedback is quick, such as in weather forecasting, comparisons with the observed values provides a basis for evaluating the goodness of the assessors and the probabilities assigned. In addition to calibration, several other characteristics of prediction performance are useful, such as refinement or sharpness. Refinement relates to a sample of probability assignments and is defined as the degree to which the assignments are near zero or one (Murphy and Winkler 1992). A well-calibrated assessor need not be a good predictor or forecaster. If the relative rate of an event A is 30%, the assessor would be well calibrated if he always assigned a probability of A equal to 30%. The refinement would, however, be poor.

4.1.2 Heuristics and Biases

People tend to use rather primitive cognitive techniques when assigning probabilities, i.e. so-called heuristics. Heuristics for assigning probabilities are easy and intuitive ways to deal with uncertain situations. The result of using such heuristics is often that the assessor unconsciously tends to put too much weight on insignificant factors. Here are some of the most common heuristics:

- *Availability heuristic*: the assessor tends to base his probability assignment on the ease with which similar events can be retrieved from memory. Events where the assessor can easily retrieve similar events from memory are likely to be given higher probabilities of occurrence than events that are less vivid and/or completely unknown to the expert.
- *Anchoring and adjusting heuristics*: the assessor tends to choose an initial anchor. Then extreme points are assessed by adjusting away from the anchor. One of the consequences is often a low probability of extreme outcomes.
- *Representativeness heuristic*: the assessor assigns a probability by comparing his knowledge about the phenomenon with the stereotypical member of a specific category. The closer the similarity between the two, the higher the judged probability of membership in the category.

The training of the risk analyst and the expert providing input to the risk analyst should make them aware of these heuristics, as well as other problems of quantifying probabilities such as superficiality and imprecision which relates to the assessor's possible lack of feeling for numerical values. Lack of precision is particularly a problem when evaluating events on the lower part of the probability scale, typically less than 1/100. Since many applications of risk analysis deal with catastrophic events, it may be interesting to examine causal factors that are considered only theoretically possible or unlikely. To some extent the situation may be improved by applying assessment aids such as a set of standardized reference events with commonly agreed probabilities that may be compared with the event under consideration, or graphical tools like the so-called probability wheel, see French and Insua (2000). Faced with rare events, however, the expert simply has difficulties in relating his uncertainty to low probability levels and in distinguishing between numbers such as 10^{-5} and 10^{-6}.

Although all experts seem to have a probability level below which expressing uncertainty in numbers becomes difficult, this level can be improved by training. Through repeatedly facing the problem of assigning probabilities to rare but observed events (the result is not known to the analyst or expert a priori), discussing the causal factors and comparing their likelihood, the analyst or expert familiarizes themself with this way of thinking. The analyst or expert will gradually feel more comfortable with applying smaller numbers, but still training alone will hardly solve this problem. It seems we must accept that the application of probability judgement has a boundary at the lower and upper ends of the probability scale beyond which probability assignments have low confidence.

Given this fact, the challenge is to design models that minimize the number of low (high) probability events to be specified. By using event trees, for example, we can try to reduce the problem to one of specifying 'reasonable' probabilities. We refer to Section 4.2.

4.1.3 Evaluation of the Assessors

The starting point for the discussion in this section is that the risk analyst would like to specify the probability distribution $P(Y \leq y)$. This probability is a measure of uncertainty; it is not an observable quantity. No true value of $P(Y \leq y)$ exists. Consequently, we cannot draw conclusions about the correctness of this probability distribution. If the pragmatic criterion applies, i.e. the probabilities can be compared to 'objective' values, assessors can be meaningfully evaluated. For example, if an analyst predicts two failures of a system during a period of one year, and the associated uncertainty is considered negligible, this assessment and the assessor would be judged as poor if there were strong evidence showing that such systems would fail at least 10 times a year. Unfortunately, the pragmatic criterion does not often apply in a risk analysis context. Sufficient relevant data do not exist. The goodness of the probability number is then more a question of who is expressing their view, what competence they have, what methods and models they use and their information basis in general, as well as what quality assurance procedures have been adopted in planning and executing the assessment. Thus we make a clear distinction between the probability number itself, which cannot be validated, and the evaluation of that probability number. Confidence in the probability assignment process is essential. This confidence is affected by several factors:

- Gap judged by the evaluator of the probability (that could be the decision-maker), between the assessor's state of knowledge and the 'best information available'.
- The evaluator considers the best information available to be insufficient.
- Motivational aspects.
- The training of the assessor in probability assignments, and in particular how to treat heuristics and biases, superficiality and imprecision, as discussed above.

If the evaluator considers the assessor's level of information (knowledge) to be significantly lower than the best information available, he would find the results of the analysis not very informative. The evaluator will be sceptical of the assessor as an expert. Trying to use the best expertise available does not fully solve this problem since in practice there will always be time and cost constraints. Even if the analyst or expert is considered to have the best information available, there could be a confidence problem. The evaluator may judge the best information available to be insufficient, and further studies are required to give a better basis for the probability specification.

A risk analyst (an expert) may assign a probability that completely or partially reflects inappropriate motives rather than their deeply felt belief regarding a specific event's outcome. As an example, it is hard to believe that a sales representative on commission would make a completely unprejudiced judgement of two safety valves where one of them belongs to a competitor firm. Another example is an engineer that was involved in the design process and is later asked to judge the probability of failure of an item he personally recommended to be installed. The engineer claims that the item is absolutely safe and assigns a very low failure probability. The management may reject the sales representative's judgement without much consideration since they believe that inappropriate motives have influenced it. The engineer's judgement might not be rejected quite so easily since the engineer is a company expert in this area. On the other hand, incentives are present that might affect the engineer's probability specification.

Motivational aspects will always be an important part of evaluating probabilities and therefore the usefulness of analyses that include expert judgements. In general, we should be aware of any incentives that in some cases could significantly affect the assignments.

4.1.4 Standardization and Consensus

When conducting many risk analyses within for example a company, there is a need for standardization of some of the probabilities to be used in the analysis, perhaps related to the distribution of the time to failure of a unit, to reduce the analysis work and ensure consistency. Such a standardization requires consensus among the various assessors in the company. In general, consensus on probabilities is usually what we desire. It is not always possible to obtain, as analysts may have different views. But when consensus can be established, it gives a stronger message.

4.2 MODELLING

This section looks at how to establish a deterministic function g such that we can write $Y = g(X_1, X_2, \ldots, X_n)$ for some observable quantities $X_1, X_2, \ldots, X_n$. Chapter 3 presented several examples of such models and we will briefly review some of them. Then we will reflect on the modelling process in general; what is the purpose of the modelling and how do we think when developing a suitable model.

4.2.1 Examples of Models

In Section 3.2.1, p. 55, we studied a cost risk model where the total cost Y was written as the sum of a number of cost elements X_i, $i = 1, 2, \ldots, k$, i.e.

$$Y = \sum_{i=1}^{k} X_i.$$

Thus the function g is simply equal to the sum of its elements or components. In this case we can quickly conclude that this is a good model as it reflects the real world accurately, provided that we have been able to include the key cost elements.

The models established in the production risk example of Section 3.2.2, p. 55, are more complex. For example, the downtime in an interval $[0, t]$, Y_t, is expressed by

$$Y_t = \int_0^t (1 - X_s) \, ds,$$

where X_s is the state process of the system, which is 1 if the system is functioning and 0 otherwise at time s. As shown in Section 3.2.2, we can write X_s as a function of the lifetimes and downtimes of the system. Thus Y_t is linked to observable quantities on a more detailed level through a deterministic function. Again we find that the model should be a good representation of the real-world system, as the system we are modelling would necessarily alternate between being up or down.

We will also make some comments on the event tree example in Section 3.3, p. 60. The model is given by the event tree shown in Figure 2.1. Clearly this is a rather rough model as it specifies for example two fatalities in the case of an explosion scenario and one in the fire scenario. In real life we could obviously have situations where these scenarios give a different number of fatalities, for instance no fatalities in the fire scenario. A possible extension of the model would be to allow the numbers of fatalities to be unknown (observable) quantities and assess associated uncertainties. This extension would give a more precise description of the real world, but the original model is simpler and it could be judged sufficiently accurate for its purpose as long as the main features of the phenomenon are reflected in the model.

Finally, we look at a case where the aim is to predict a distribution function $F(t)$ in a setting where we can define an appropriate population of similar units. We may think of $F(t)$ as the proportion of units with lifetimes less than or equal to t. Therefore, in this setting, $F(t)$ is an observable quantity and we can apply the principles of Chapter 3. As a model of $F(t)$ we introduce for example the exponential distribution with parameter λ, such that we can write $F(t) = F(t|\lambda) = 1 - \exp\{-\lambda t\}$. Note that in this case the parameter is an observable quantity, representing a state of the world; it is the average number of failures per unit of exposure time for the whole population of units. Letting H denote an uncertainty distribution of λ, the distribution of $F(t)$, for a fixed

t, takes the form

$$P(F(t) \le x) = \int_{\{\lambda : F(t|\lambda) \le x\}} dH(\lambda).$$

Furthermore, we can calculate for example a 90% prediction interval curve for the function F by

$$P(F(\cdot|\lambda_1) < F \le F(\cdot|\lambda_2)) = \int_{\lambda_1}^{\lambda_2} dH(\lambda) = 0.90,$$

where λ_1 are λ_2 are the 5% and 95% quantiles for H.

4.2.2 Discussion

A model is a simplified representation of a real-world system. The general objective of developing and applying models in this context is to arrive at risk measures based on information about related quantities and a simplified representation of real-world phenomena. Detailed modelling is required to identify critical factors contributing to risk and evaluate the effect of risk-reducing measures. The simplified form makes models suitable for analysis, and in model construction this property is traded off against the need for complexity that produces sufficiently detailed results. Typical factors governing the selection of models are the form of the detailed system information and its level, the resources available in the specific study and whether the focus is on the overall risk level or on comparing decision alternatives. In general the advances seen within computer technology have improved the conditions for analysing complex models.

Since models are used to reflect the real world, they only include descriptions of relationships between observable quantities. Probabilistic expressions reflect uncertainty or lack of knowledge related to the values of such quantities. Modelling is a tool that allows us to express our uncertainty in the format found most appropriate to fulfil the objectives of performing the analysis.

Experience data applied in risk analysis are often given in the form of the number of occurrences of an outcome y out of a number of trials n, registered during similar activity in the past. However, the 'similar activity' often comprises a mix of experiences resulting in data representing an average system. This makes it hard to differentiate between the decision alternatives at hand. It becomes especially hard to defend alternatives that involve new technology not represented in the data.

In most cases the data do not reflect system-specific information, e.g. related to local operating conditions, and technical and organizational measures already implemented. Such additional system information usually exists as a mix of detailed system specifications and expert knowledge. To be able to reflect such information, further system modelling is required. Differentiation between the decision alternatives is achieved through a more detailed system representation. Referring to the set-up above, this implies identification of factors (quantities) $\mathbf{X}$ to be included in the model $Y = g(\mathbf{X})$.

The problem of small probability numbers can often be avoided by modelling. Assume that we are interested in quantifying our uncertainty related to whether the event A will occur in a given period as input to a risk analysis. If A is judged by the experts to be improbable and the experts have difficulties in relating to it quantitatively, the problem may be handled by shifting the focus to observable quantities on a lower causal level, associated with a higher probability level. For example, if A is judged dependent on the occurrence of conditions B and C, the expert may express his uncertainty with respect to these events instead, and a probability of A, $P(A)$, may be assigned by $P(A) = P(B)P(C|B)$. Another alternative is to formulate A by a limit state function, i.e. A occurs if $g(\mathbf{X}) < 0$, where g is a limit state function (Section 2.1.3). The probability $P(A)$ can then be specified by expressing uncertainty about the event through the probability distributions of the observable quantities $\mathbf{X}$.

In summary, we can say that under our predictive approach to risk and risk analysis, modelling is a tool for identifying and expressing uncertainty, hence it is also a means for potentially reducing uncertainty. The uncertainty can be identified by including more system-specific information in the analyses, in terms of an expanded information basis for uncertainty statements and in terms of the model structure itself. Furthermore, modelling adds flexibility to the risk analyses since it allows us to express uncertainty in the format found most appropriate to obtain the objectives of the analysis.

A topic closely related to the use of models, and widely discussed in the literature, is model uncertainty. Several approaches to interpretation and quantification of model uncertainty are proposed in the literature, see Section 2.1.3 and Bibliographic notes of the present chapter. In our setting, a model $Y = g(\mathbf{X})$ is a purely deterministic representation of factors judged essential by the analyst. It provides a framework for mapping uncertainty about the observable quantity of interest, Y, from expressions of epistemic uncertainty related to the observable quantities, $\mathbf{X}$, and does not in itself introduce additional uncertainty. In this setting, the model is merely a tool judged useful for expressing knowledge about the system. The model is part of the background information on the probability distribution specified for Y. If we change the model, we change the background information.

It is not relevant to talk about uncertainty of a model. What is interesting to address is the goodness or appropriateness of a specific model to be used in a specific risk analysis and decision context. Clearly, a model can be more or less good in describing the world. No model reflects all aspects of the world, but it should reflect key features. We return to this topic in Section 4.4.3.

4.3 ASSESSING UNCERTAINTY OF Y

The problem is to specify a probability distribution $P(Y \leq y)$ for $y \geq 0$, given a background information K represented as observational data (hard data) $y_1, y_2, \ldots, y_n$ and as expert knowledge. These hard data could be more or less relevant. Now, how should we proceed to specify $P(Y \leq y)$? Several approaches can be used:

- derivation of an assigned distribution based on classical statistics;
- analyst judgement using all sources of information;
- formal expert elicitation.

These approaches are discussed in more detail in Sections 4.3.1 to 4.3.3.

The Bayesian approach gives a unified approach to the specification of $P(Y \leq y)$. To apply this approach, the common procedure is to introduce a parameter, say θ, representing a state of nature, such that we can write

$$P(Y \leq y) = \int P(Y \leq y|\theta) \, dH(\theta), \qquad (4.1)$$

where H is the prior distribution of θ and $P(Y \leq y|\theta)$ is normally given by a common parametric distribution function, for example the exponential. Bayes' theorem tells us how to update the prior distribution when new data becomes available to obtain a posterior distribution. The Bayesian approach will be presented in more detail in Section 4.3.4; see also Appendix A. Here we consider when to use a full Bayesian approach with the specification of a prior distribution and apply equation (4.1), instead of a more direct assignment process for determining $P(Y \leq y)$, such as the three approaches referred to above. Note that these three approaches may also be viewed as Bayesian, although they are largely based on direct probability assignments without introducing a parameter; see the discussion on page 79.

4.3.1 Assignments Based on Classical Statistical Methods

Consider first the problem of specifying the probability that $Y = 1$ in the case that Y is a binary quantity (indicator function) taking the values 0 or 1. Then direct use of classical statistics would lead to the probability assignment

$$P(Y = 1) = \frac{1}{n} \sum_{i=1}^{n} y_i, \qquad (4.2)$$

i.e. $P(Y = 1)$ is given as the relative portion of 'successes' of the n observations $y_1, y_2, \ldots, y_n$. So, for example, if we have 3 successes out of 10 observations, we obtain $P(Y = 1) = 0.3$. This is our (i.e. the analyst's) assessment of uncertainty related to the value of Y.

In this framework $P(Y = 1)$ is specified according to equation (4.2); it is not an estimate of an underlying true probability $P(Y = 1)$ as in the classical setting, but an assessment of uncertainty related to the occurrence of $Y = 1$. Thus, for the above example, $P(Y = 1) = 0.3$, whereas in the classical setting, $P^* = 0.3$, where P^* is an estimate of $P(Y = 1)$, i.e. $P(Y = 1) \approx 0.3$, hopefully.

This method is appropriate when the analyst judges the observational data to be relevant for the uncertainty assessment of Y, and the number of observations

n is large. What is considered sufficiently large depends on the setting. As a general guidance, we find that about 10 observations is enough in many cases to specify the probabilities using this method, provided that not all observations are either 1 or 0. In this case the classical statistical procedure gives a probability equal to 1 or 0, which we would normally not find adequate for expressing our uncertainty about Y. Other procedures then have to be used; see the next two sections.

Now, suppose that Y takes values in the set of real numbers, and as above we assume that the analyst judges the observational data to be relevant for the uncertainty assessment of Y, and the number of observations n is large. Then we can proceed along the same lines as for the binary case but we specify $P(Y \leq y)$ by the equation

$$P(Y \leq y) = \frac{1}{n} \sum_{i=1}^{n} I(y_i \leq y), \tag{4.3}$$

where I is the indicator function, which is 1 if the argument is true and 0 otherwise. Thus $P(Y \leq y)$ is given by the empirical distribution function in the classical statistical set-up.

In most cases we would prefer to use a continuous function for $P(Y \leq y)$, as it is mathematically convenient. Such a function is obtained by a fitting procedure where the empirical distribution is approximated by a continuous function, for example a normal distribution function. Classical statistical methods for fitting a distribution function to observed data are the natural candidate for this procedure, see Appendix A.2. As for the binary case, note that we use classical inference merely as a tool for assessing our uncertainty distribution for Y, not for estimating an underlying true distribution function for Y.

This procedure works with a large number of observations, but what if n is not large, say 6, or what if most of the observations are zero, say, and we are most concerned about a possible large value of Y, i.e. the tail of our uncertainty distribution of Y? Or what if the data are not considered sufficiently relevant? Clearly, in these cases it is problematic to use the above procedure, as the information given by the data is so limited. Other procedures should then be adopted.

4.3.2 Analyst Judgements Using All Sources of Information

This is a method commonly adopted when data are absent or are only partially relevant to the assessment endpoint. A number of uncertain exposure and risk assessment situations are in this category. The responsibility for summarising the state of knowledge, producing the written rationale, and specifying the probability distribution rests with the analyst. It is very likely that two different analysts will produce two different descriptions of the present state of knowledge and probability distributions.

Now, how does the analyst derive one particular probability distribution? Consider first the binary case, where the problem is to specify $P(Y = 1)$. The starting point is that the analyst is experienced in assigning probabilities expressing uncertainty, so he has a number of references points; the analyst has a feeling for what 0.5 means in contrast to 0.1, for example. A probability of 0.1 means that the analyst's uncertainty related to the occurrence of $Y = 1$ is the same as when drawing a favourable ball from an urn with 10% favourable balls under standard experimental conditions. To facilitate the specification, the analyst may also think of some type of replication of similar events as generating $Y = 1$, and think of the probability as corresponding to the proportion of 'successes' that they would predict among these events. Suppose the analyst predicts 1 success out of 10, then they would assign a probability 0.1 to $P(Y = 1)$. Note that this type of reasoning does not mean that the analyst presumes the existence of a true probability, it is just a tool for simplifying the specification of the probability.

Now consider the general case of assessing the distribution of Y when the possible value of Y is on the real line. The simplest approach is to specify probabilities as above for the events $Y \leq y_i$ or $Y > y_i$, for suitable numbers y_i. Often one starts with a percentage, say 90%, and then specifies the value y such that $P(Y > y) = 0.90$. Combining such quantile assessments with a specified distribution class, such as the normal distribution or a lognormal distribution, only a few assessments are needed (typically two, corresponding to the number of parameters of the distribution class).

An alternative approach for the specification of $P(Y \leq y)$ is to use the maximum entropy principle, see p. 83.

To specify the probability distribution, the analyst may consult experts in the subject of interest, but the uncertainty assessment is not a formal expert elicitation as explained below.

4.3.3 Formal Expert Elicitation

This approach requires the analyst to identify and bring together individuals acknowledged as experts in the subject of concern. Here is a typical procedure. The analyst trains the experts in the assessment problem and disseminates among the experts all relevant information and data. The experts are then required to formalize and document their rationales. They are interviewed and asked to defend their rationales before committing to any specific probability distribution. The experts specify their own distribution by determining quantiles.

Sometimes weights are assigned to the experts to distinguish differences in expertise. Some argue that the selection of high-quality experts at the outset is mandatory and that all experts used for the final elicitation should be given the same weight. Others argue that the experts should be given the opportunity to assign weights to themselves.

Formal approaches to expert elicitation seemingly place all responsibility for quantifying the state of knowledge on the panel of experts. The method is extremely difficult to rebuke, except by conducting new experiments on the uncertain quantity of interest or convening a separate independent panel of experts.

It is a basic principle of our approach to risk analysis that the analyst is ultimately responsible for the assessment, and as such, the analyst is obliged to make the final call on the probability distribution. Experts have advanced knowledge in rather narrow disciplines and are unlikely to devote the time necessary (even with training) to become as familiar as the analyst with the unique demands of the assessment question. However changing the experts' distributions should not be done if this possibility is not a part of an agreed procedure for elicitation between the analyst and the experts.

We recommend that formal expert elicitation is undertaken when little relevant data can be made available and when it is likely that the judgement of the analyst will be subject to scrutiny, perhaps resulting in costly project delays. Formal expert elicitation could be very expensive, so it requires adequate justification.

Experts may specify their own probability distributions, or they could provide the analyst with information for him or her to process and finally transform to a probability distribution. This latter approach has the advantage that the experts can speak their own language and avoid the somewhat abstract formalism of using probabilities. On the other hand, it may be difficult for the analyst to fully understand the expert judgements if they are just reports of knowledge, with no reference to the probability scale.

Building consensus, or rational consensus, is of major concern when using expert opinions. Five principles are often highlighted (Cooke 1991);

- *Reproducibility*: it must be possible to reproduce all calculations.
- *Accountability*: the basis for the probabilities assigned must be identified.
- *Empirical control*: the probability assignments must in principle be susceptible to empirical control.
- *Neutrality*: the methods for combining or evaluating expert opinion should encourage experts to state their true opinions.
- *Fairness*: all experts are treated equally, prior to processing the results of observations.

We find these principles appropriate, but a remark on empirical control as stated in Section 4.1 is in place. Empirical control does not apply to the probability at the time of assignment. When conducting a risk analysis we cannot verify an assigned probability, as it expresses the analyst's uncertainty prior to observation.

4.3.4 Bayesian Analysis

To illustrate the Bayesian thinking, here are three examples. Other examples are presented in Section 4.4.

Health risk

Suppose we test a patient when there are indications that they have a blood disease. Let X be 1 or 0 according to whether the test gives positive or negative response. Furthermore, let θ be the true condition of the patient, the state of nature, which is defined as 2 if the patient is seriously ill, 1 if the patient is moderately ill, and 0 if the patient is not ill at all. From general health statistics, suppose that 2% of the relevant population is seriously ill, 10% is moderately ill, and 88% is not ill at all from this disease.

From these health statistics and without using additional information about the patient, we can specify a prior distribution

$$P(\theta = 2) = 0.02, \quad P(\theta = 1) = 0.10, \quad P(\theta = 0) = 0.88. \quad (4.4)$$

Now suppose we know from experience that the test will give a positive response in 90% of the cases if it is being applied on a patient that is seriously ill. If the patient is moderately ill, the test will give a positive response in 60% of the cases, whereas if the patient is not ill, the test will give a false response in 10% of the cases. From this information we can formulate the following conditional probabilities:

$$P(X = 1|\theta = 2) = 0.90,$$
$$P(X = 1|\theta = 1) = 0.60,$$
$$P(X = 1|\theta = 0) = 0.10.$$

We refer to this as the likelihood function $L(\theta)$. Combining these probabilities and those given by (4.4), we can compute the posterior probability $P(\theta = 2|X = 1)$, i.e. the probability that the patient is serious ill given that the test gives a positive response. Simple probability calculus gives

$$P(X = 1) = P(X = 1|\theta = 2)P(\theta = 2) + P(X = 1|\theta = 1)P(\theta = 1)$$
$$+ P(X = 1|\theta = 0)P(\theta = 0)$$
$$= 0.90 \times 0.02 + 0.60 \times 0.10 + 0.10 \times 0.88$$
$$= 0.166,$$

and using Bayes' theorem;

$$P(\theta = 2|X = 1) = \frac{P(X = 1|\theta = 2)P(\theta = 2)}{P(X = 1)}$$
$$= \frac{0.90 \times 0.02}{0.166} = 0.11.$$

More generally, we may write the conditional distribution of θ given $X = x$, which is called the posterior distribution of θ, as

$$f(\theta|x) = \frac{L(\theta)f(\theta)}{f(x)},$$

where f is used as a generic symbol to express a distribution. Thus the posterior distribution $f(\theta|x)$ is proportional to $L(\theta)f(\theta)$.

The calculations have produced a probability of 0.11 that the patient is seriously ill given that the test has shown a positive response. This is a rather low number, which the doctor needs to take into consideration when communicating with the patient. Some would say that the test should not be used at all as it is simply too poor.

However, the situation can be improved by performing an additional test to provide more information. This corresponds to an A test and a B test in a doping context. We would like to compute the probability for the patient to be seriously ill given that both tests have shown a positive response. Let X_i be 1 or 0 according to whether the test i gives positive or negative response, $i = 1, 2$. The sought probability can then be written as $P(\theta = 2|X_1 = 1, X_2 = 1)$.

Consider first the situation after the first test has been performed and the test has given a positive response. Instead of using the uncertainty distribution of θ based on the health statistics, we now start with the updated probabilities (the posterior distribution) $P(\theta = 2|X_1 = 1)$, $P(\theta = 1|X_1 = 1)$ and $P(\theta = 0|X_1 = 1)$ based on the information that the first test showed a positive response. Using Bayes' theorem we established above that $P(\theta = 2|X_1 = 1) = 0.11$. Similarly, we find that

$$P(\theta = 1|X_1 = 1) = \frac{0.60 \times 0.10}{0.90 \times 0.02 + 0.60 \times 0.10 + 0.10 \times 0.88} = 0.36,$$

$$P(\theta = 0|X_1 = 1) = \frac{0.10 \times 0.88}{0.90 \times 0.02 + 0.60 \times 0.10 + 0.10 \times 0.88} = 0.53.$$

In this example we view the tests as conditionally independent in the sense that the probability that the second test gives a positive response given that the patient is seriously ill (moderately ill, not ill), does not depend on the result of the first test. Thus we have

$$P(X_2 = 1|\theta = 2, X_1 = 1) = P(X_2 = 1|\theta = 2) = 0.90,$$

$$P(X_2 = 1|\theta = 1, X_1 = 1) = P(X_2 = 1|\theta = 1) = 0.60,$$

$$P(X_2 = 1|\theta = 0, X_1 = 1) = P(X_2 = 1|\theta = 0) = 0.10,$$

which are the same probabilities used for the calculations of $P(\theta = 2|X = 1)$ above.

Hence we replace $P(\theta = i)$ by $P(\theta = i|X_1 = 1)$ and apply Bayes' theorem to obtain

$$P(\theta = 2|X_1 = 1, X_2 = 1) = \frac{0.90 \times 0.11}{0.90 \times 0.11 + 0.60 \times 0.36 + 0.10 \times 0.53} = 0.27.$$

This posterior probability is much better than 0.11, but still it is rather low.

The calculations demonstrate how Bayes' theorem is used to update probabilities when new information becomes available. Note that the probability calculus above is general in the sense that it also applies to a classical interpretation of probability as so far we have used relative frequencies as the basis for our probability numbers. Now we would like to go one step forward and include specific information that the doctor has about the condition of the patient. Suppose that the patient has shown some rather strong symptoms of being seriously ill. The doctor finds that the probability distribution (4.4) is not reflecting his view concerning the state of the patient, given the present state of knowledge. Instead the doctor assigns the following probabilities expressing his uncertainty about θ:

$$P(\theta = 2) = 0.40, \quad P(\theta = 1) = 0.40, \quad P(\theta = 0) = 0.20. \quad (4.5)$$

From this starting point, the probability calculations are similar to those shown above and they lead to a probability of 0.58 that the patient is seriously ill, given that the first test has shown a positive response. And if both tests show a positive response, the sought probability is found to be 0.69.

This Bayesian analysis provides the probabilities of interest given the observations. This is in contrast to classical statistical hypothesis testing where the probabilities of interest are computed prior observations, given the parameter θ. To be more specific, we can formulate a null hypothesis H_0 and an alternative hypothesis H_1 by

$$H_0: \theta = 0 \text{ and } H_1: \theta > 0,$$

i.e. we test whether the patient is ill, starting from the null hypothesis that he is not ill. We reject the null hypothesis and claim that the patient is ill if both tests give positive results. The significance level of the test is 1% as

$$P(X_1 = 1, X_2 = 1 | \theta = 0) = 0.10 \times 0.10 = 0.01,$$

given the above assumptions. In the Bayesian analysis we would compute $P(\theta > 0 | X_1 = 1, X_2 = 1)$, i.e. the probability that the patient is ill given that both tests give positive results. We find this probability is equal to 99.6%.

The Bayesian analysis provides a recipe to calculate the posterior distribution $P(\theta = i | X_1 = x_1, X_2 = x_2)$, the probability of the parameter being a specific value, given the observations and the background information. This distribution is a complete description of our understanding of θ. There is nothing more to be said. Summing over $i = 1$ and $i = 2$, this distribution provides our entire understanding of whether H_1 is true.

Criminal law

The defendant in a court of law is either truly guilty G or not guilty $\overline{G}$. The guilt is uncertain and we describe this uncertainty by a probability $P(G)$. It is convenient to work in terms of odds:

$$o(G) = P(G)/P(\overline{G}).$$

If we have data available in the form of evidence B, we update probabilities according to Bayes' formula, yielding

$$o(G|B) = \frac{P(B|G)}{P(B|\overline{G})} o(G),$$

involving multiplication of the original odds by a likelihood ratio expressing our probabilities of the data given the state of the world G or $\overline{G}$. As the trial proceeds, further evidence is introduced and successive multiplications by the likelihood ratios determine the final odds.

This type of calculation could be used as a basis for a judgement of guilty or not guilty. We may also think of the jury communicating the final odds $o(G|B)$, where B is the totality of all admitted evidence.

Accident risk

Let us return to the event tree example in Section 3.3, p. 60, and let us reconsider the problem of specifying the probability of ignition, $P(A)$, and the distribution of the number of leakages, X, occurring in a one-year period.

Following the presentation of Section 3.3, the uncertainty assessment of the ignition event is in accordance with the approach in Section 4.3.2. The question now is how to perform a Bayesian analysis according to (4.1), p. 72. And what are the possible benefits of adopting this analysis compared to the more direct approach?

Adopting a full Bayesian analysis, the first step would be to introduce a parameter. In this case it would be p, interpreted as the proportion of times ignition will occur when considering an infinite or very large number of similar situations to the one analysed. If we knew p, we would assign a probability of A equal to p, i.e. $P(A|p) = p$. Hence from (4.1) we obtain

$$P(A) = \int p \, dH(p), \qquad (4.6)$$

where H is the prior distribution of p. Now how should we interpret (4.6)?

The standard Bayesian framework and its interpretation go as follows. To specify the probabilities related to A, a direct assignment could be used, based on everything we know. Since this knowledge is often complex, of high dimension, and much in the background information may be irrelevant to A, this approach is often replaced by the use of probability models, which is a way of abridging the background information so that it is manageable. Probability models play a key role in the Bayesian approach. In this case the probability model is simply $P(A|p) = p$, where p is the parameter of the probability model. The parameter p is also known as a chance – it is an objective property of the constructed sequence or population of situations. It is not a probability for the assessor, though were p known to the assessor, it would be the assessor's probability of A, or any event of the sequence. The parameter p is unknown and our uncertainty related to its value is specified through a prior distribution $H(p)$. Later we will return to the problem of specifying the prior distribution H. We see from equation (4.6), that the unconditional distribution of A is simply given by the mean in the prior distribution of p. Note that both $P(A)$ and H are specified given the background information. Thus the uncertainty distribution of A is expressed via two probability distributions, p and H. The two distributions reflect what is commonly known as aleatory (stochastic) uncertainty and epistemic (state of knowledge) uncertainty.

This framework is based on the idea that there exists, or there can be constructed through a thought experiment, a sequence of events A_i related to 'similar' situations to the one analysed. The precise mathematical term used to define what is similar is 'exchangeability'. Random quantities $X_1, X_2, \ldots, X_n$ are judged exchangeable if their joint probability distribution is invariant under permutations of coordinates, i.e.

$$F(x_1, x_2, \ldots, x_n) = F(x_{r_1}, x_{r_2}, \ldots, x_{r_n}),$$

where F is a generic joint cumulative distribution function for $X_1, X_2, \ldots, X_n$ and equality holds for all permutation vectors $(r_1, r_2, \ldots, r_n)$, obtained by switching (permuting) the indices $\{1, 2, \ldots, n\}$; see Appendix A, p. 156. Exchangeability means a judgement about indifference between the random quantities. It is a weaker requirement than independence because, in general, exchangeable random quantities are dependent.

In our case we may view the random quantities as binary, i.e., they take either the value 0 or 1, and if we consider an infinite number of such quantities, judged exchangeable, then it is a well-known result from Bayesian theory that the probability that k out of n are 1 is necessarily of the form

$$P\left(\sum_{i=1}^{n} X_i = k\right) = \binom{n}{k} \int_0^1 p^k (1-p)^{n-k} dH(p), \tag{4.7}$$

for some distribution H. This is a famous result and is known as de Finetti's representation theorem. Thus, we can think of the uncertainties (beliefs) about observable quantities as being constructed from a parametric model, where the random quantities can be viewed as independent, given the parameter, together with a prior distribution for the parameter. The parameter p is interpreted as the long-run frequency of 1s. Note that it is the assessor that judges the sequence to be exchangeable, and only when that is done does the frequency limit exist for the assessor.

Bayesian statistics is mainly concerned with inference about parameters of the probability models. Starting from the prior distribution H, this distribution is updated to a posterior distribution using Bayes' theorem; see the health example above and Appendix A.

We see that the Bayesian approach as presented above allows for fictional parameters based on thought experiments. These parameters are introduced and their uncertainty is assessed.

In our view, applying the standard Bayesian procedures gives too much focus on fictional parameters, established through thought experiments, see the discussion in Section 2.3.2. The focus should be on observable quantities. A rewriting of the standard Bayesian presentation is thus required, to establish a theory consistent with our predictive approach.

For this example we would use a simple direct approach as presented in Section 4.3.2. Direct probability assignments should be seen as a useful supplement to establishing probability models where we need to specify prior distributions of parameters. We may use parametric distribution classes, but we should be careful about interpretation. We return to this topic below, following the examination of a somewhat more complex case: assessing the uncertainty of X, the number of leakages occurring in one year.

Suppose we have observations $x_1, x_2, \ldots, x_n$ related to previous years, and let us assume that these data are considered relevant for the year studied. We would like to predict X. How should we do this? The data allow a prediction simply by using the mean $\bar{x}$ of the observations $x_1, x_2, \ldots, x_n$. But what about uncertainties? How should we express the uncertainty? Suppose the observations $x_1, x_2, \ldots, x_n$ are 4, 2, 6, 3, 5, so that $n = 5$ and the observed mean is equal to 4. In this case we have rather strong background information, and we suggest using the Poisson distribution with mean 4 as our uncertainty distribution of X. For an applied risk analyst, this would be the natural choice as the Poisson distribution is commonly used for event type analysis and the historical mean is 4. Now, how can this uncertainty distribution be justified? Well, if this distribution reflects our uncertainty about X, it *is* justified, and there is nothing more to say. This is a subjective probability distribution and there is no need for further justification. But is a Poisson distribution with mean 4 reasonable, given the background information? We note that this distribution has a variance not larger than 4. By using this distribution, 99% of the mass is in values less than 10.

Adopting the standard Bayesian thinking, as outlined above, using the Poisson distribution with mean 4, means that we have no uncertainty about the parameter λ, which is interpreted as the long-run average number of failures when considering an infinite number of exchangeable random quantities, representing similar systems as the one being analysed. According to the Bayesian theory, ignoring the uncertainty about λ, gives misleading overprecise inference statements about X, see Bernardo and Smith (1994: 483). This reasoning is valid if we work within a setting where we are considering an infinite number of exchangeable random quantities. In our case, however, we just have one X, so what do we gain by making a reference to limiting quantities of a sequence of similar hypothetical Xs? The point is that given the observations $x_1, x_2, \ldots, x_5$, the choice of the Poisson distribution with mean 4 is in fact reasonable under certain conditions on the uncertainty assessments. Consider the following argument. Suppose that we divide the year $[0, T]$ into time periods of length T/k, where k is for example 1000. Then we may ignore the possibility of having two events occurring in one time period, and we assign an event probability of $4/k$ for the first time period, as we predict 4 events in the whole interval $[0, T]$. Suppose that we have observations related to $i - 1$ time periods. Then for the next time period we should take these observations into account – using independence means ignoring available information. A natural way of balancing the prior information and the observations is to assign an event probability of $(d_i + 4n)/((i - 1) + nk)$, where d_i is equal to the total number of events that occurred in $\left[0, T(i - 1)/k\right]$, i.e. we assign a probability equal to the total number of events occurred per unit of time. It turns out that this assignment process gives an approximate Poisson distribution for X. This can be shown for example by using Monte Carlo simulation. The Poisson distribution is justified as long as the background information dominates the uncertainty assessment of the number of events occurring in a time period. Thus from a practical viewpoint, there is no problem in using the Poisson distribution with mean 4. The above reasoning provides a justification of the Poisson distribution, even with not more than one or two years of observations.

Now consider a case with no historical data. Then we will probably find the direct use of the Poisson distribution as described above to have too small a variance. The natural approach is then to implement a full parametric Bayesian procedure. But how should we interpret the various elements of the set-up? Consider the following interpretation.

The Poisson probability distribution $p(x|\lambda)$ is a candidate for our subjective probability for the event $X = x$, and $H(\lambda)$ is a confidence measure, reflecting for a given value of λ the confidence we have in $p(x|\lambda)$ being able to predict X. If we have several X_i, similar to X, and λ is our choice, we believe that about $p(x|\lambda) \times 100\%$ of the X_i will take a value equal to x, and $H(\lambda)$ reflects for a given value of λ, the confidence we have in $p(x|\lambda)$ being able to predict the number of X_i taking the value x. We refer to this as the confidence interpretation.

Following this interpretation, we avoid the reference to a hypothetical infinite sequence of exchangeable random quantities. We do not refer to $H(\lambda)$ as an uncertainty distribution as λ is not an observable quantity.

If a suitable infinite (or large) population of 'similar units' can be defined, in which X and the X_i belong, then the above standard Bayesian framework applies as the parameter λ represents a state of the world, an observable quantity. Then $H(\lambda)$ is a measure of uncertainty and $p(x|\lambda)$ is truly a model – a representation of the proportion of units in the population having the property that the number of failures is equal to x. We may refer to the variation in this population, modelled by $p(x|\lambda)$, as aleatory uncertainty, but still the uncertainty related to the values of the X_i is seen as a result of lack of knowledge, i.e. the uncertainty is epistemic.

The same type of thinking can be used for the uncertainty assessment of the ignition event A. The confidence interpretation would in this case be as follows. Our starting point is that we consider alternative values p for expressing our uncertainty about A. The confidence we have in p being able to predict A is reflected by the confidence distribution H. If we have several A_i, similar to A, and p is our choice, we believe that about $p \times 100\%$ of the A_i would occur, and $H(p)$ reflects for a given value of p, the confidence we have in p being able to predict the number of A_i occurring.

The above analysis provides a tool for predicting the observable quantities and assessing associated uncertainties. When we have little data available, modelling is required to get insights and hopefully reduce our uncertainties, see Section 4.2. The modelling also makes it possible to see the effects of changes in the system and to identify risk contributors.

Specifying the prior distribution

Our starting point is the fundamental equation of Bayesian analysis:

$$P(Y \leq y) = \int P(Y \leq y|\theta) \, dH(\theta),$$

where H is the prior distribution of θ. Prior distributions should reflect the knowledge possessed before the relevant data are at hand. This is the Bayesian standpoint. However, in practice the specification of the prior is often difficult and certain classes of technique are being used. One of these is the use of so-called non-informative priors. The idea is to specify a distribution reflecting total lack of information about the parameter. For example, in the binomial case a non-informative prior distribution for the parameter p is given by the uniform distribution on the interval $[0, 1]$. But what about situations where the parameter takes values on $[0, \infty)$? Should we use a so-called improper prior having a density equal to 1 for all parameter values?

No, such a distribution should be avoided, and even when the non-informative distribution is proper, we should avoid it. We believe that in most practical cases the analyst would have some knowledge, and that information should be incorporated, to give a proper informative probability distribution. Consider the Poisson distribution example above and the problem of specifying a prior distribution for λ. We could ignore values of λ that are very large, so why should we then use a prior that gives positive weight to such values?

Probably the choice of non-informative priors is more motivated from the need of having an 'objective' prior, rather than reflecting total lack of knowledge. Non-informative distributions may be a simple way of establishing consensus, but it could mean ignoring significant information.

The use of so-called conjugate distributions is another principle frequently adopted. When using such distributions, the prior and posterior distribution belong to the same distribution class. For example, the Poisson and gamma distributions are conjugate. Adopting this principle makes it relatively simple to carry out Bayesian updating, i.e. to establish the posterior distribution. But if the prior does not reflect your opinion, it should not be used.

An interesting approach for specifying the prior distribution is to use a maximum entropy prior. This approach means specification of some features of the distribution, for example the mean and the variance, but not the whole distribution. Then a mathematical procedure gives a distribution with these features and in a certain sense, minimum information beyond that. Refer to Bedford and Cooke (2001: 73) for the details.

If θ is an observable quantity, the specification of the uncertainty distribution of θ is similar to the one discussed for Y above and in the following section. Illustrations are given above for the health and accident risk examples.

4.4 UNCERTAINTY ASSESSMENTS OF A VECTOR X

In this section several examples discuss how to assess uncertainties and specify probability distributions for a set of observable quantities.

4.4.1 Cost Risk

We refer to the cost risk problem introduced in Section 3.2.1, p. 52. We have established a model

$$Y = \sum_{i=1}^{k} X_i,$$

where Y represents the investment cost related to a project and the X_i, $i = 1, 2, \ldots, k$, represent more detailed cost elements. If we judge these cost elements to be independent, we can use the procedures of the previous section to assess uncertainties and specify probability distributions for each X_i and by probability calculus or Monte Carlo simulation establish the distribution for Y.

In practice the use of independence is often problematic. We may for example think of a situation where the cost elements are all strongly influenced by the oil price, and the question is then how to incorporate this in the assessments. Alternative approaches can be used; the following one is based on remodelling and is one of the simplest.

Let X be the value of an underlying factor, for example the oil price, influencing the cost elements X_i. It is common to refer to X as a latent quantity (variable). We write $X_i(X)$ to show the dependency of X. Given X, we judge the cost elements to be independent. Then by specifying an uncertainty distribution of X, and of X_i given X, we can compute the uncertainty distribution of Y. By Monte Carlo simulation this is rather easy to do. We draw a number x from the distribution of X, and then use this as a starting point for drawing values of $X_i(x)$. These data are then used to produce a Y value. The same procedure is repeated until we obtain the resulting probability distribution of Y.

The challenge is to find a simple way of expressing the judged dependencies. In the example above where the X_i are related to a quantity X, we may go one step forward and express X_i for example by the equation

$$X_i = a_i X + b_i + X_i', \tag{4.8}$$

where the observable quantities X_i' and X are judged independent, and X_i' has a distribution $F_{X_i'}$, with mean 0 and variance τ_i^2. By (4.8) the influence of the factor X on X_i has been explicitly described through remodelling, such that independence of the adjusted quantities $X_i - a_i X$ can be justified. It follows that

$$Y = \left(\sum_i a_i\right) X + \sum_i b_i + \sum_i X_i',$$

and this distribution can rather easily be found, for example by Monte Carlo simulation, as all unknown quantities on the right-hand side of the equality sign, are judged independent. The basis for using equation (4.8) would normally be a regression analysis. The idea is to plot (using a so-called scatter plot) observations (x, x_i) of (X, X_i) in a two-dimensional diagram and fit the data to a line adopting standard least squares linear regression; see Appendix A.2.4.

Another way of incorporating dependency is to specify correlation coefficients ρ_{ij} between X_i and X_j. To interpret these coefficients, we consider our uncertainty distribution of the pairs X_i, X_j. From these distributions various summarizing measures can be derived, including the correlation coefficient defined by

$$\rho_{ij} = E(X_i - \mu_i)(X_j - \mu_j)/\sigma_i \sigma_j,$$

HOW TO ASSESS UNCERTAINTIES AND SPECIFY PROBABILITIES

where μ_i and σ_i are the mean and standard deviation of X_i, respectively. In practice we assign values for ρ_{ij} without specifying the simultaneous distribution of X_i and X_j. If the simultaneous distribution of the cost elements is a multivariate normal distribution with parameters μ_i, σ_i and ρ_{ij}, see Appendix A.1.5, then Y also has a normal distribution, with mean

$$EY = \sum_{i=1}^{k} \mu_i$$

and variance

$$\mathrm{Var}\, Y = \sum_{i=1}^{k} \sigma_i^2 + 2 \sum_{i<j} \rho_{ij}\sigma_i\sigma_j.$$

An example with $n = 2$ is presented in Section 2.2.2. Thus the task is to specify the expected values, the standard deviations and correlation coefficients. If we have available a large amount of relevant data, we can use the empirical counterparts as a basis for assigning values for these quantities. We see that by using normal distributions, the mathematics become simple.

We may establish the expected value and the variance of Y by the above formulas without specifying the uncertainty distributions of the observable quantities X_i. Together the mean and variance provide measures of uncertainty and risk. But this way of thinking does not produce an uncertainty distribution of Y, and that is our objective. Using normal distributions we have seen that for establishing the joint distribution of the X_i, it is sufficient to specify the marginal distribution for each uncertain quantity X_i and the correlation coefficients of each pair of the X_i. Using some transformations of the marginal distributions, we can generalize this result. It is not straightforward as we need to specify correlation coefficients of these transformations, not the correlation coefficients of X_i and X_j. Refer to Bedford and Cooke (2001: 329) for the details.

An interesting alternative approach for specifying the joint distribution is presented in Bedford and Cooke (1999); see also Bedford and Cooke (2001). It is based on the specification of the marginal distributions, as well as probabilities of the form $P(X_1 > x_1 | X_2 > x_2)$, where x_1 and x_2 are the 50% quantiles of the distributions of X_1 and X_2, respectively. Using a mathematical procedure, a minimal informative distribution is establish based on this input. A minimal informative distribution is in a sense the most 'independent' joint distribution with the required properties.

4.4.2 Production Risk

In the production risk example studied in Section 3.2.2, p. 55, we assigned distributions for the uptimes and downtimes of components being repaired or replaced at failure when considering observations in a time period $[0, t]$. The consecutive component lifetimes and repair times are denoted T_{im} and R_{im}, respectively, where i refers to the ith component. These quantities are unknown and we express our uncertainty related to what will be the true values by probability distributions.

The question is now how to assess these uncertainties and specify the probability distributions. Ideally, a simultaneous distribution for all lifetimes and repair times should be provided, but this is not feasible in practice. So we need to simplify. Let time t be a fixed point in time. Suppose we have strong background information concerning the component lifetimes and the repair times. Then as a simplification of the uncertainty assessments, we could judge all T_{im} and R_{im} to be independent and use the same distribution F_i for all lifetimes and the same distribution G_i for all repair times of component i. This was done in Section 3.2.2. It is a rather strong simplification; we ignore learning when conditioned on the values of some of the lifetimes and repair times. But as discussed in the Poisson example, in some cases the background information is such that we could justify the use of independence. Suppose for example that we use exponentially distributed lifetimes and fixed repair times. Then we can argue, along the same lines as for the Poisson example, p. 81, that the Poisson process is reasonable to use when considering operational time (we ignore the downtimes), with the parameter λ, the expected number of failures per unit time, given by the observed mean. In the general case we would use a full Bayesian analysis.

Now, how should we perform the full Bayesian analysis? We first establish a class of probability distributions for the lifetimes and repair times. To simplify, suppose that we use fixed repair times and exponentially distributed lifetimes with parameter λ_i. Then if Y denotes the performance measure being studied, we can write

$$P(Y \leq y) = \int P(Y \leq y | \lambda) \, dH(\lambda),$$

where λ is the vector of the λ_i and H is the prior distribution of λ. Given λ, the distribution of Y is found by using that the lifetimes are independent with exponential distributions having parameters λ_i; we are back to the independent case. So it remains to establish the prior distribution H. Refer to Section 4.3.4, p. 82. If we include uncertainty related to repair times and use for example a Weibull distribution to describe the lifetimes, the analysis will be similar, but more complicated when it comes to the specification of the prior distribution.

4.4.3 Reliability Analysis

Traditional reliability analysis

We use the standard reliability nomenclature introduced in Section 2.1.3. As a simple example, let us consider a parallel system of two components. The state of the system, the model of the world, is given by the monotone structure function

$$\Phi = \Phi(\mathbf{X}) = 1 - (1 - X_1)(1 - X_2),$$

where $\mathbf{X} = (X_1, X_2)$, is the vector of binary component states. The task is to determine our unreliability $P(X_1 = 0, X_2 = 0)$.

Now, probability calculus gives in general

$$P(X_1 = 0, X_2 = 0) = P(X_2 = 0 | X_1 = 0) P(X_1 = 0),$$

and if we judge X_1 and X_2 to be independent, we have

$$P(X_1 = 0, X_2 = 0) = P(X_2 = 0) P(X_1 = 0).$$

Thus, by specifying the probabilities $P(X_1 = 0)$ and $P(X_2 = 0 | X_1 = 0)$, we arrive at the unreliability. The marginal probability $P(X_1 = 0)$ is often rather easy to specify as we have performance data for the components, but it is more difficult to specify the conditional probability $P(X_2 = 0 | X_1 = 0)$ as we seldom have available data for this conditional situation. So what do we do then? Well, we can make a direct assignment of the probability expressing uncertainty about $X_2 = 0$ given $X_1 = 0$, but in most cases it would be more attractive to model the dependency. One way of doing this is to identify the source causing the dependency – the common cause – and specify the proportion of failures due to this common cause. Let $X = 0$ denote the event that this common cause occurs. Then we obtain

$$P(X_1 = 0, X_2 = 0) = P(X_1 = 0, X_2 = 0 | X = 0) P(X = 0)$$
$$+ P(X_1 = 0, X_2 = 0 | X \neq 0) P(X \neq 0)$$
$$\approx 1 \times P(X = 0) + P(X_2 = 0) P(X_1 = 0) P(X \neq 0),$$

and we are back to the independent case. Note that when assigning $P(X_i = 0)$ in the above equation we should reflect that these probabilities are in fact conditional on the non-occurrence of the common cause. We notice that this example is analogous to the introduction of the X in the cost risk analysis example.

Now, suppose that the components are of the same type, for example two similar machines. We have sampled these two machines from a huge stock of similar machines. If we have strong background information, we would put $P(X_2 = 0 | X_1 = 0) = P(X_2 = 0)$, i.e. judge $X_1 = 0$ and $X_2 = 0$ independent, as the information that $X_1 = 0$ would not add much information relative to the information already available.

Next, suppose that we have no information whatsoever about the performance of this type of machine and would like to assign a probability for the system to be functioning at a specific point in time. What would then be our unreliability $P(X_1 = 0, X_2 = 0)$?

As we have no knowledge about the performance of this type of machine, we would assign a failure probability of 0.5 for a machine, i.e. $P(X_1 = 0) = P(X_2 = 0) = 0.5$. There should be no discussion about this. If we judge X_1 and X_2 independent, we are through, as that would give a system unreliability of $0.5 \times 0.5 = 0.25$. But given $X_1 = 0$, we should change (increase) our probability of $X_2 = 0$ as we have received information about the performance of this type of machine. How should we incorporate this in our analysis?

Let p be the proportion of machines that are not functioning, out of the large population of similar machines. We note that p is an observable quantity. If we have no information about the performance of this type of machine, we could specify a uniform prior distribution over the interval [0, 1] to express our uncertainty about the value of p. If we knew the value of p, we would assign a probability of component i not functioning equal to p, and judge the components independent. Thus we have

$$P(X_1 = 0, X_2 = 0) = \int_0^1 (p^2 \times 1)\,dp = \frac{1}{3}. \qquad (4.9)$$

We see that the unreliability of the system is 1/3, which is higher than 1/4 obtained by judging X_1 and X_2 to be independent.

Although this is a rather theoretical case – we are seldom in a situation with no information – it is illustrative, showing the importance of precise understanding of information, observable quantities, uncertainties and probabilities.

Now, suppose that the machines we are studying are in a specific operational and maintenance environment, such that we cannot refer to a population of similar machines. We just have a few relevant observable quantities, including X_1 and X_2. How should we then proceed?

With no information, we would use equation (4.9). But the interpretation is different, as we have no population of similar machines to refer to. We do not introduce fictional populations and quantities (parameters). We consider different values of p for describing our uncertainty about the occurrence of $X_i = 0$. The confidence we have in p for being able to predict the X_i is reflected by a uniform distribution. For example, specifying $p = 0.1$ means that we would be fairly sure that no machine failures occur. If $p = 0.6$, we would predict one machine failure out of the two. The confidence we have in the various p being able to predict the number of X_i is reflected by the uniform distribution.

Using a beta prior distribution with parameters α and β, the resulting predictive distribution of Y, the number of components functioning, i.e. $Y = X_1 + X_2$, has a beta-binomial distribution with parameters $(2, \alpha, \beta)$. Thus we may specify a prior beta distribution and then derive the predictive distribution, or we could make a direct assignment of the parameters of the beta-binomial distribution. This latter approach means a stronger focus on the observable quantities X_i and Y, but would probably be more difficult to carry out in practice.

To determine X_i it may in some cases be appropriate to relate the functioning or not functioning to a lifetime T_i, such that $X_i = 1$ if $T_i > t$, where t is the time of interest. Then we may specify our uncertainty related to the value of X_i by specifying a probability distribution for T_i, for example an exponential distribution $1 - e^{-\lambda_i t}$, where λ_i is the failure rate of the component, given by $\lambda_i = 1/ET_i$. This distribution expresses our uncertainty about the value of T_i. See Section 4.4.2 for a discussion of how to use and interpret such a class of distribution functions.

A sensitivity analysis can be used in this setting by changing the input probabilities, in most cases to the extremes, meaning that we compute system reliability given that component i has zero reliability. In this way we identify the importance of the component reliability and the improvement potential related to improvement of this component. An alternative approach that is also used for importance identification is to look for the effect of small changes: How quickly does the system reliability index change when the input probability changes? The measure is specified by taking the partial derivative of the probability index with respect to the probability; it is known as Birnbaum's measure.

Structural reliability analysis

Let us reconsider the load strength example in Section 2.1.3, where the limit state function is given by $Y = g(\mathbf{X}) = X_1 - X_2$. Here X_1 represents a strength measurement of the system and X_2 represents a load measurement. Thus g is the model and by expressing uncertainties of (X_1, X_2) using a density function f, system reliability can be expressed as

$$P(Y < 0) = \int_{\{\mathbf{x}:\ g(\mathbf{x}) < 0\}} f(\mathbf{x})\, d\mathbf{x}.$$

One such distribution f could be the multivariate (bivariate) normal distribution with parameters μ_i, σ_i^2 and ρ. Suppose we have strong background information about the values of X_i, for example corresponding to 20 observations x_{ij} that are all considered relevant for X_i. Then we may use fixed values of the parameters μ_i, σ_i^2 and ρ.

Now suppose that we do not have such background information and we would like to update our predictions and uncertainty assessments when new data become available. Then we would adopt the full Bayesian procedure with specification of a prior distribution on the parameters μ_i, σ_i^2 and ρ. Mathematically this leads to the same formulas as used in the classical approach with uncertainty analysis; see equations (2.5) and (2.6). But there are some important differences.

There exist no true values of the parameters, unless they are observable quantities. In that case the prior (and posterior) distribution is an uncertainty distribution. In the general case, the prior (and posterior) distribution expresses our confidence in the parameter values being able to predict the X_i; see the discussion above on page 82.

It is not relevant to speak about modelling uncertainty, but the 'goodness' of the models to represent the world. The model is a part of the background information, and is reported along with the assigned probabilities. Let us discuss this a little further using the load strength model as an illustration. Let Y be the true rest capacity of the system at the time of interest, when taking into account the load. Using the model $g(\mathbf{X}) = X_1 - X_2$, we have put $Y = g(\mathbf{X})$. This means a simplification, and in SRA it is common to introduce an error term X_0, say, such that we can write $Y = X_0(X_1 - X_2)$. This gives a better model, a more accurate description of the world. As a simplification, we judge

X_0 and $X_1 - X_2$ to be independent; our uncertainty about the ratio between the true capacity and the measured capacity $X_0 = Y/(X_1 - X_2)$ is not influenced by the value of the capacity indicator $X_1 - X_2$. This simplification should be supported by observations of the true capacity and the measured capacity for comparable situations. Thus by specifying uncertainty distributions for X_0 and $X_1 - X_2$, we arrive at an uncertainty distribution of Y. If we use the means as predictors, the true capacity is predicted by $EX_0 \cdot (EX_1 - EX_2)$. By introducing X_0, the uncertainty in Y increases. In SRA applications the explanation of this is model uncertainty. In our setting, there is no such thing as model uncertainty. If we use the model $X_1 - X_2$ to express uncertainty about the true capacity Y, this means that we have conditioned on the use of this model. If we find that the model $X_1 - X_2$ is not sufficiently accurate for its purpose, we should improve the model. Using the equation $Y = X_0(X_1 - X_2)$ gives an accurate model, but to express uncertainties in this case, we need to simplify and use independence, which is a rather strong simplification. In this particular situation it may be acceptable, but in other cases it would not be acceptable. Furthermore, often it may be difficult to find relevant data to support the uncertainty analysis of X_0. We perform the analysis as we have little information about Y. If we had a strong database for Y, we could make a direct assignment of the distribution of Y, and there would be no need for the modelling.

4.5 DISCUSSION AND CONCLUSIONS

Reliability of probabilities has been thoroughly discussed in the literature. Many researchers link probability assessment and utility; they find it hard to devise a reliable form of measurement for uncertainty assessments that is separate from utility considerations. We disagree. We have to acknowledge that the standard measurement criteria cannot be met. Coherence applies, that is all. Linking probability with utility does not solve the problem, it just disturbs and confuses the assessor. In most cases we prefer to see the uncertainty assessments as a separate process providing a basis for the decision-making, see Chapter 5. Scoring rules and empirical control in general aim to train assessors and compare them if it is possible to obtain relevant feedback, as in meteorology where one is concerned about repetitions of a single type of event, like 'rain tomorrow'. In most other areas, however, this feedback is not available.

Scoring rules are also motivated by the desire to provide incentives for the predictors to honestly report their probabilities; see Cooke (1991), de Finetti (1962: 359) and Winkler (1996b).

If two persons have the same background information, would that mean that they have the same uncertainties, and therefore the same probabilities? No, in our setting 'the probability' does not exist – probability is an expression by a person based on some knowledge about an observable quantity. Often we would experience similar numbers if the knowledge is about the same, but there are no formal constraints on the framework implying that my judgement should be the same as yours if we have the same knowledge. A probability is a judgement, and there is no strict mechanical procedure producing one correct value.

Assessing a probability $P(Y \leq y)$ (given the background information) directly can be viewed as a basic procedure of the Bayesian paradigm, see Lindley (2000: 304). According to the standard Bayesian thinking, there is however a better way to proceed, to study the mechanisms that operate, linking Y and other states of the world. This means the introduction of a probability model with parameters, say θ, such that we have

$$P(Y \leq y) = \int P(Y \leq y|\theta) \, dH(\theta);$$

see equation (4.1) p. 72. Yes, this is a fundamental approach of the Bayesian thinking, and of ours, but care should be taken when introducing these type of models, as discussed in this chapter, over when to introduce them and how to interpret the various elements. Basically, there are two ways of applying this type of modelling: by restricting θ to observable quantities, or allowing fictional parameters related to thought-constructed long-run behaviour, expressed as parameters of probability distribution classes. In our framework we have highlighted the former way of thinking; see the examples in this chapter and Chapter 3. In the health risk example we introduced the state of the world θ expressing the health condition of the patient, and in the accident risk example we introduced an event tree model with states of the world given by the number of leakages X, and the events A and B. Using such modelling is often a better way to proceed than direct assignments of $P(Y \leq y)$ – it is easier to perform coherent judgements and hopefully we obtain better predictions. But we have avoided the introduction of the latter category of modelling, based on fictional parameters. We do not introduce an uncertainty distribution over the limiting proportion of events of type A, for example, and assess uncertainties of such fictional quantities. It does not contribute to a better understanding of the processes generating the data – rather it means the creation of uncertainty, which we need not consider.

We have tried to advise on the modelling process; the theory is available, but how should we use it in a practical context? Simplifications and approximations are needed.

Our basic idea that there is only one type of uncertainty is sometimes questioned. It is felt that some probabilities are easy to assign and feel sure about, others are vague and it is doubtful that the single number means anything. Should not the vagueness be specified? To provide a basis for the reply, let us look at an example. A coin is thrown and the event A denotes that it shows heads. In another example, we test a drug and the event B denotes that the drug is better than the old with a particular pair of patients (the meaning of 'better' is well defined and is not an issue here). In the absence of any information about the type of coin, we would assign a probability of A equal to 1/2, and this probability is firm in that we would almost all be happy with it. With the drug test we would have an open mind about its effectiveness and similarly ascribe a probability of B equal to 1/2. This latter value of 1/2 is vague and one does not feel so sure about it as with the coin. It seems that we have a firm, objective probability of 1/2 and one vague, subjective probability of 1/2.

The reply puts focus on the background information of the probabilities and the available knowledge to be used as a basis for assessing the uncertainties.

We know more about the process leading to a head in coin tossing than in the drug example. If we consider 1000 throws, we would be quite sure that the proportion of heads, which we denote p, would be close to 1/2. Most people would assign very low probabilities for observing say less than 100 heads. In the drug example we would, when considering 1000 pairs of patients, have less information about the result, i.e. q, representing the portion of the 1000 pairs of patients benefiting more from the new drug than the old. The new drug could be a complete flop and the old cure is vastly to be preferred, meaning that we would assign a rather high probability also to low values of q. Both low and high values of q are much more probable than low and high values of p, simply because we know that coins could not easily be that biased, whereas drugs could well be quite different. These different probabilities reflect vagueness and firmness that are respectively associated in our minds with the original probabilities. In the coin example, the background information is so strong that observations would not easily change our assessment, whereas in the drug example, medical evidence would probably lead us to believe in the effectiveness of the new drug. This can be shown formally using Bayes' theorem for updating probabilities.

This example demonstrates the importance of paying attention to appropriate performance measures. In the above example it is not A and B, but p and q. When evaluating probabilities in a decision-making context, we always need to address the background information, as it provides a basis for the evaluation.

Many people are alarmed, in particular in scientific matters, by using probabilities as a subjective measure of uncertainty as we do. The approach is seen to be in conflict with science, which searches for objective statements. Our view is that complete knowledge about the world does not exist in most cases, and we provide a tool for dealing with these uncertainties based on coherence. If sufficient data become available, consensus may be achieved, but not necessarily as there are always subjective elements involved in the assessment process. The objective truth when facing future performance does not exist.

BIBLIOGRAPHIC NOTES

The three types of criteria considered in Section 4.1 – pragmatic, semantic (calibration) and syntactic – are discussed in Lindley *et al.* (1979). Calibration and the use of scoring rules are reviewed and discussed by Cooke (1991), Winkler (1996b), Lindley (1982), among others. Some key references to the theory on heuristics are Cooke (1991), Kahneman *et al.* (1982), Otway and Winterfeldt (1992), Tversky and Kahnemann (1974) and Kahneman *et al.* (1982).

For other papers on the 'goodness' of probability assignments, see Berg Andersen *et al.* (1997) and Winkler (1968, 1986).

Section 4.2 is based on the ideas of the predictive paradigm presented in Chapter 3, and we refer to Apeland *et al.* (2002) and Nilsen and Aven (2003). Modelling uncertainty is discussed by Dewooght (1998), Draper (1995), Zio and Apostolakis (1996) and Nilsen and Aven (2003).

The discussion of probability assignments in Sections 4.3.1 to 4.3.3 is based on Hoffman and Kaplan (1999). We also refer to Apeland *et al.* (2002). Much literature exists on the Bayesian approach. Good reference books and papers are Lindley (1978, 1985, 2000), Bedford and Cooke (2001), Barlow (1998), Bernardo and Smith (1994), Singpurwalla (1988, 2002) and Singpurwalla and Wilson (1999). The health risk example of Section 4.3.4 is taken from Natvig (1997). The accident risk example is based on Aven (2001).

As an alternative to the presented approach for establishing the Poisson approximation, we could study the predictive distribution of X in a full Bayesian analysis, assuming that $x_1, x_2, \ldots, x_5$ are observations coming from a Poisson distribution, given the mean λ and using a suitable (e.g. non-informative) prior distribution on λ. Restricting attention to observable quantities only, a procedure specified in Barlow (1998: Ch. 3) can be used. This procedure, in which the multinomial distribution is used to establish the Poisson distribution, is based on exact calculation of the conditional probability distribution of the number of events in sub-intervals, given the observed number of events for the whole interval.

Note that for our direct assignment procedure using the k time periods, the observations $x_1, x_2, \ldots, x_5$ are considered a part of the background information, meaning that this procedure does not involve any modelling of these data. In contrast, the more standard Bayesian approach requires that we model $x_1, x_2, \ldots, x_5$ as observations coming from a Poisson distribution, given the mean λ.

Overviews of the problem of specifying prior distributions are given by Singpurwalla and Wilson (1999) and Bedford and Cooke (2001). See also Lindley (1978, 2000), Bernardo and Smith (1994) and Vose (2000).

The reliability analysis of two components in a Bayesian setting is a classical illustration of the importance of information when specifying probabilities. It is discussed by Bedford and Cooke (2001), among others.

The question of whether two persons with the same background information would assign the same probabilities, is discussed in Lindley (2000: 302). The discussion on the vagueness and firmness of probabilities is based on Lindley (1985: 112).

The importance of making a sharp distinction between uncertainty assessment and utility has been emphasized by many researchers; see for example Good (1950, 1983). The point they are making is that subjective probability assignments need not necessarily always reveal themselves through choice. Probability expresses uncertainty, and usually through intervals of upper and lower probabilities rather than single numerical values. Intervals may be useful in some situations for expressing subjective probabilities. For example, when the probabilities are very low, or during an early stage of an assignment process, we consider a set of probabilities to express our uncertainty. But as a general principle we search for single numerical values. That means a drive for information and knowledge, and the right focus, namely our uncertainty about the observable quantities, and not the lack of ability to express this uncertainty.

5

How to Use Risk Analysis to Support Decision-Making

This chapter considers how to use risk analysis in a decision-making context when adopting the predictive approach to risk and uncertainty presented in Chapters 3 and 4. The purpose of risk analysis is to support decision-making, not to produce numbers. It is from this starting point we have established our predictive approach to risk and uncertainty, and in this chapter we will see how we can fit this framework into a more general decision-making setting. If the purpose of risk analysis is to support decision-making, that is, help the decision-maker to make decisions, we need some idea of what a good decision is. Decision-making is of course not about making decisions, but about making *good* decisions. Therefore, we first, in Section 5.1 address the fundamental issue of what is a good decision. There is no simple answer and there are several different views. In this chapter we review the issue and give some guidelines on how we should plan for obtaining good decisions. It discusses the link between risk analyses and formal decision analyses, such as cost-benefit analyses and Bayesian decision analyses. We see the need for a structure for how to apply risk analysis in a decision-making context and we establish some principles that may be useful in practice. Several examples in Section 5.2 discuss the implementation of these principles. Two classification schemes for risk problems are presented in Section 5.3. These schemes are used to discuss the need for risk and uncertainty analyses, formal decision analyses as well as risk management policies.

The presentation is *prescriptive* in that it aims to describe good principles and methods that should be used to select a course of action in practice. It is closely linked to *normative* approaches, such as the expected utility paradigm, which is a norm or a standard on how a person ought to behave based on a logical study of choice between decisions within a mathematical framework. The exposition is not *descriptive* in the sense of describing how people actually make decisions. However, when establishing the principles and methods, we have of course examined descriptive theory and results reported in the literature and used our experience from real life. The aim is to establish a structure for decision-making that produces good decisions, or improved decisions, defined in a suitable way, based on a realistic view of how people can act in practice.

This book discusses the use of risk analysis as a tool for decision-making, and it touches on aspects of risk treatment, risk acceptance and risk communication. Risk treatment is the process and implementation of measures to modify risk, including measures to avoid, reduce (optimize), transfer or retain risk. Risk transfer means sharing with another party the benefit or loss associated with a risk. It is typically effected through insurance. It is, however, beyond the scope of this book to discuss in detail all aspects of risk management, i.e. all coordinated activities to direct and control an organization with regard to risk. The many challenges for an organization related to defining objectives, to avoid, reduce, transfer and retain risks we just briefly look into. The various disciplines and application areas need to define their own risk management system, tailored to the specific situations of interest.

5.1 WHAT IS A GOOD DECISION?

Consider someone contemplating an investment in a stock of 1 million dollars for a period of one year. At the end of that period, the stock may be worth more or less than the original sum spent on purchasing it. If the person does not invest in this stock, he will leave the money in the bank. So the decision alternatives are invest or leave in the bank. Suppose that at the end of the one-year period the stock has a value of $(1 + X)$ million dollars, and in the case the person leaves the money in the bank, $1 + Y$. Now what would be a good decision?

Well, the immediate, natural answer would be the alternative that gives the best outcome. In this case we compare the bank interest rate Y and the increase (decrease) in the stock value, X. At the end of the one-year period we can observe which decision is the best by simply looking at the outcomes, and money provides the obvious scale of preference.

In most decision-making situations, however, we do not have a simple scale of preference, and we are not able to observe the outcomes. If we compare alternative concepts for the development of an offshore oil and gas field, how do we measure the goodness of the outcomes? A number of factors are relevant, including costs, environmental and safety issues, reputation, and political issues such as employment. We refer to these factors as the attributes of the problem. We would not be able to make observations for more than one of the alternatives, as the decision will exclude all but one. For the chosen alternative we can see how it performs, but changes may have been implemented so that the alternative in operation is significantly different from the one defined at the decision point.

We see that using the outcomes as a basis for judging the goodness of a decision is problematic; it cannot be done at all in most cases. Yet this outcome-centred thinking is important, in our view, as it makes us have a clear focus on what the objectives and preferences are. The problem is, however, that this thinking does not help us very much in making good decisions. The decisions are made prior to observing the outcomes. What is a good decision when *contemplating* various alternatives? Factors such as costs, safety, reputation, politics, etc., are still relevant, but we do not know for sure the possible outcomes. We need to make decisions under uncertainty, and the challenge is to establish some guidelines on how to do this such that we make a good decision.

5.1.1 Features of a Decision-Making Model

There are two basic ways of thinking to reach a good decision:

(i) Establish an optimization model of the decision-making process and choose the alternative which maximizes (minimizes) some specified criteria.
(ii) See decision-making as a process with formal risk and decision analyses to provide decision support, followed by an informal managerial judgement and review process resulting in a decision.

This book adopts approach (ii) as an overall structure, meaning that we see decision analysis strictly as an aid for decisions. This does not mean that we cannot see examples where approach (i) is appropriate, but considering varying degrees of the managerial judgement and review process, we may think of approach (i) as a special case of approach (ii). Regardless of the approach, we will not be able to avoid the fact that some decisions will be followed by negative outcomes. But by following a decision-making process in line with the principles in (ii), we would expect that a collective of decisions will produce overall positive outcomes in relation to the objectives, when seen together.

Figure 5.1 shows the main features of this way of thinking about decision-making. The starting point is a decision problem and often this is formulated as a task of choosing among a set of decision alternatives. Let us use the alternative concepts for the development of a gas and oil field as an example. At this stage of the development project, the management has at hand a number of possible alternatives. The problem is to identify one or two for further detailing and optimization. Much has already been decided when a set of alternatives to be further evaluated has been defined. Suppose we decided at an early stage to adopt a well-proven technology. Then we would exclude cases that require new technology. In a practical setting, the number of alternatives to be evaluated needs to be manageable, therefore many alternatives could be excluded at an early stage when uncertainties are large. Further studies might have shown that these alternatives are favourable compared to those being evaluated. The set of alternatives is typically defined through an integrated process involving experts and managers. The experts would often specify an initial list as a basis for discussion. The development of alternatives would largely be driven by the boundary conditions of the decision problem, as judged by the experts and management. The boundary conditions include stakeholders' values, for example formulated as organizational goals, criteria, standards and preferences, as well as views expressed by politicians, environmentalists and others. Experts and managers have a background, values, preferences, etc., that could significantly influence the selection process of alternatives. We have to appreciate

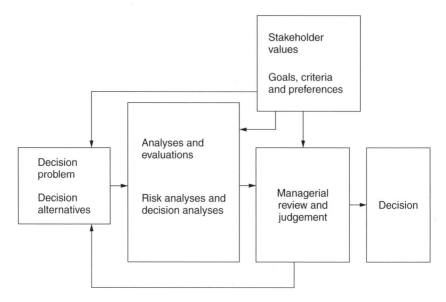

Figure 5.1 Basic structure of the decision-making process

the subjective element in this creative part of decision-making, establishing an appropriate set of alternatives. We know that people have personal agendas, but by ensuring that the process involves a sufficiently broad group of personnel, the generated alternatives should provide the necessary basis for identifying a good alternative.

5.1.2 Decision-Support Tools

Now, suppose we have given a set of decision alternatives. Before management makes a decision, it needs some support as a basis for its decision. It needs to know more about the consequences of choosing one alternative instead of another. Risk analysis provides such decision support. It gives predictions of the performance of the various alternatives with related uncertainty assessments. This information is then linked to other attribute assessments. Consider again the oil and gas development project presented earlier. The idea is to analyse and evaluate factors such as investment costs, operational costs, market deliveries and regularity, technology, safety and environmental issues, and political aspects. For costs, market deliveries, safety and environmental issues, quantitative analyses are conducted in line with our predictive framework. For other important aspects such as the political, only qualitative analyses and evaluations would normally be performed. The total of these analyses and evaluations is used as a decision basis. We refer to this as a multi-attribute analysis. Such an analysis provides structuring and overview of the problem – it provides useful insights. Before a decision is made, management reviews and evaluates the decision-support information, and relates it to values formulated as goals, criteria and preferences. There is no strict procedure on how to perform this managerial process. It is an individual process based on the constraints of the structure in Figure 5.1. This structure gives a prescription of how to conduct the managerial process in practice, and in this sense the process is documented and traceable. It is a prescription, but it is not very detailed and specific. Note that the structure model given by Figure 5.1 does not show all the feedback processes, for example the managerial review and judgement may result in modified analyses.

The managerial process means trade-offs of a number of attributes. These trade-offs could be made more or less explicit. Let us look at a simple example.

Two alternatives are compared, A and B. Associated with alternative A there is a gain of 0.2 (i.e. a cost -0.2) and associated with alternative B there is a gain of 0.1 (i.e. a cost -0.1). We may think of one cost unit as 1 million dollars. The assigned probabilities of a fatality for the two alternatives are 2/100 and 1/100, respectively. These probabilities are associated with a time period of 10 years, say. We assume for the sake of simplicity that there are no other factors to consider. What alternative should be chosen? How should we balance cost and safety? In general it is not possible to answer this question. Balancing cost and safety is a management task, which is based on goals, criteria and preferences, but in most cases there is no direct line from these to a specific decision. Alternative A means a reduced cost compared to B, but a higher probability of a fatality. What is the value of a probability of a fatality of 1/100 compared to 2/100? Again we cannot give a general answer, but we could compute an index, a cost-effectiveness index, expressing cost per expected life saved, which gives a reference and a link between the two dimensions cost and safety. We see that the index in this case is $0.1/[(2/100) - (1/100)]$, which is equal to 10. The reasoning is as follows. To go from alternative A to alternative B it would cost 0.1, and the expected number of saved lives would be $2/100 - 1/100$. Then if we find an index of 10 (million dollars) as too high to be justified, the analysis would rank alternative A before alternative B.

A number of studies have been conducted to measure implicit values of a statistical life. The costs differ dramatically, from net savings to costs of nearly 100 billion dollars. In industry it is common to use a reference value in the area $1 - 20$ million dollars.

Another way of performing this type of analysis is to express a cost value for a statistical life, that is, the expected cost per expected saved life. Suppose that we assign a value of 2 to such a cost. Then the total statistical expected 'gain' associated with alternative A would be

$$0.2 - 2 \times 2/100 = 0.16,$$

whereas for alternative B, the corresponding value would be

$$0.1 - 2 \times 1/100 = 0.08.$$

The conclusion would thus be that alternative A is preferable as the expected gain is 0.16 compared to a gain of 0.08 for alternative B. In practice we need to take into account time and the discounting of cash flow, but the above calculations show the main principles of this way of balancing cost and benefit. It

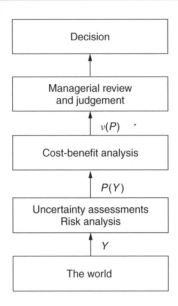

Figure 5.2 Basic structure of the decision-making process when a cost-benefit analysis is used

is common to refer to this type of analysis as a cost-benefit analysis. Note that we use a decision aid interpretation of the cost-benefit analysis, which means that the analysis is just a tool for providing insight before making the decision. There are no objective values of the analysis. This is in contrast to one common interpretation of cost-benefit analysis, searching for objective prices and probabilities; see the discussion below and the bibliographic notes.

Figure 5.2 shows the structure of the decision-making process when we use a cost-benefit analysis as described above. The starting point is the world and observable quantities Y, representing for example costs or number of fatalities. Risk and uncertainty analyses are conducted producing probabilities, denoted $P(Y)$. In our example the P values for the number of fatalities are given by $1/100$ and $1/200$. Based on these analyses, a cost-benefit analysis is carried out resulting in performance measures $v(P)$, for example expected cost per expected saved life, or expected NPV. These measures, which are based on the probabilistic quantities established in the uncertainty assessments, are reviewed and a decision is made.

In this book we focus on tools for decision-making, and cost-benefit analysis as described above is just an example of such a tool. It provides input to the decision-maker, not the decision. By presenting the results of the analysis as a function of the value of a statistical life, we can demonstrate the sensitivity of the analysis conclusions. We should acknowledge that decisions need to be based on managerial review and judgement. The decision-support analyses need to be evaluated in light of the premises, assumptions and limitations of these analyses. The analyses are based on background information that must be reviewed together with the results of the analyses. Considerations need to be given to factors such as:

- the decision alternatives being analysed;
- the performance measures analysed;
- the fact that the results of the analyses represent judgements;
- the difficulty of assessing values for costs and benefit, and uncertainties;
- the fact that the analysis results apply to models, i.e. simplifications of the world, and not the world itself.

The weight that the decision-maker will put on the results of the analyses depends on the confidence he has in the analyses and the analysts. Here are some important issues: Who are the analysts? What competence do they have? What methods and models do they use? What is their information basis in general? What quality assurance procedures have they adopted in the planning and execution of the analyses? Are the analysts influenced by some motivational aspects? These are the same types of issue as we discussed when evaluating the goodness of probability assignments, see Section 4.1.3.

In our setting the analysis provides decision support, not hard recommendations. Thus we may for example consider different values of a statistical life, to get insight into the decision. Searching for a correct objective number is meaningless, as no such number exists; the statistical life used in the analysis is a value that represents an attitude to risk and uncertainties and that attitude may vary and depend on the context. When using a one-dimensional scale, uncertainties of observable quantities are mixed with value statements about how to weigh the different assessed uncertainties. Then we cannot expect to obtain consensus about the recommendations provided by the cost-benefit analysis as there are always different opinions about how to look at risk in a society. Adopting a traditional cost-benefit analysis, an alternative with a low expected cost is preferred to an alternative with a rather high value, even if the latter alternative would mean that we can ignore a probability of a serious hazard, whereas this cannot be done in the former case. In traditional cost-benefit analysis it is also common to discount the values of statistical lives, and often this means that negligible weight is put on consequences affecting future generations. It is of paramount importance that the cost-benefit analyses are reviewed and evaluated, as we cannot replace difficult ethical and political deliberations with a mathematical one-dimensional formula, integrating complex value judgements.

Another approach for performing the trade-offs between the attributes is to carry out a Bayesian decision analysis.

Formal Bayesian decision analysis: maximization of expected utility

The cost-benefit analysis approach requires balancing various assessed uncertainties – costs and accident risk in our example – not costs and number of fatalities, as required when using a Bayesian utility approach. In our example the possible consequences for the two alternatives are $(2, X)$ and $(1, X)$, where the first component of $(\cdot, \cdot)$ represents the benefit and X represents the number of fatalities, which is either 1 or 0. Now, what is the utility value of each of these consequences? Well, the best alternative would obviously be (2, 0), so let us give this consequence the utility value 1. The worst consequence would be (1, 1), so let us give this consequence the utility value 0. It remains to assign utility values to the consequences (2, 1) and (1, 0). Consider balls in an urn with u being the proportion of balls that are white. Let a ball be drawn at random; if the ball is white, the consequence (2, 0) results, otherwise the consequence is (1, 1). We refer to this lottery as '(2, 0) with a chance of u'. How does '(2, 0) with a chance of u' compare to achieving the consequences (1, 0) with certainty? If $u = 1$ it is clearly better than (1, 0), if $u = 0$ it is worse. If u increases, the gamble gets better. Hence there must be a value of u such that you are indifferent between '(2, 0) with a chance of u' and a certain (1, 0), call this number u_0. Were $u > u_0$ the urn gamble would improve and be better than (1, 0); with $u < u_0$ it would be worse. This value u_0 is the utility value of the consequence (1, 0). Similarly, we assign a value to (2, 1), say u_1. As a numerical example we may think of $u_0 = 90/100$ and $u_1 = 1/10$, reflecting that we consider a life to have a high value relative to the gain difference. Now, according to the utility-based approach, a decision maximizing the expected utility should be chosen.

For this simple example, we see that the expected utility for alternative A is equal to

$$1 \times P(X = 0) + u_1 \times P(X = 1) = 1.0 \frac{98}{100} + 0.1 \frac{2}{100} = 0.982,$$

whereas for alternative B we have

$$u_0 \times P(X = 0) + 0 \times P(X = 1) = 0.9 \frac{99}{100} + 0 \frac{1}{100} = 0.891.$$

The conclusion is that alternative A is to be preferred. Observe that the expected values computed above are in fact equal to the probability of obtaining the best consequence, namely a gain of two and no fatalities. To see this, note that for alternative A, the consequence (2, 0) can be obtained in two ways, either if $X = 0$, or if $X = 1$ and we draw a white ball in the lottery. Thus by the law of total probability, the desired results follow for alternative A. Analogously we establish the result for alternative B.

We conclude that maximizing the expected gain would produce the highest probability of the consequence (2, 0) and as the alternative is the worst, (1, 1), we have established that maximizing the expected utility value gives the best decision. This is an important result. Based on requirements of consistent (coherent) comparisons for events and for consequences, we are led to the inevitability of using the expected utility as a criterion for choosing decisions among a set of alternatives.

Figure 5.3 shows the structure of the decision-making process when utilities are used. As in the cost-benefit case, the starting point is the world, represented by Y. Uncertainty assessments are conducted, i.e. risk analysis resulting in probabilities $P(Y)$, and utilities $u(Y)$ are elicited. It is a key element of this

HOW TO USE RISK ANALYSIS TO SUPPORT DECISION-MAKING

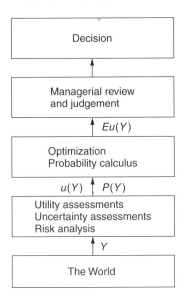

Figure 5.3 Basic structure of the decision-making process when utilities are used

approach that there is a sharp separation between uncertainty assessments and value judgements expressed by the utilities. Combining $P(Y)$ and $u(Y)$, we arrive at the expected value, $Eu(Y)$, and maximization of this measure gives the optimal decision alternative within the given framework. The decision is then made based on managerial review and judgement.

Again, the analysis would give decision support, not the decision. Managerial review and judgement are needed to produce the decision. The analysis needs to be evaluated in light of the premises, assumptions and limitations discussed earlier for the cost-benefit analyses. The expected utility approach provides knowledge and insight into the decision-making process, through the assessment process and by the use of sensitivity analyses, but it would be a management failure not to see beyond the mathematical optimization.

Furthermore, in practice, decisions need to be taken by a group of people with their own array of probabilities and utilities, but the expected utility approach is only valid for a single decision-maker. No coherent approach exists for the multiple problem; see the bibliographic notes. If the group can reach consensus on the judgements, probabilities and utilities, we are back to the single decision-maker situation. Unfortunately, life is not that simple in many cases – people have different views and preferences. Reaching a decision, then, is more about discourse and negotiations than mathematical optimization.

5.1.3 Discussion

We have looked at two approaches for aiding decision-making to balance costs and benefits: cost-benefit analysis and maximization of expected utility. Now, which approach should be taken?

The expected utility approach is attractive as it provides recommendations based on a logical basis. If a person is coherent in his preferences among consequences and his opinions about uncertainty quantities, it can be proved that the only sensible way for him to proceed is by maximizing expected utility. For a person to be coherent when speaking about the assessment of uncertainties of events, the requirement is that he follows the rules of probability. When it comes to consequences, coherence means that if c_1 is preferred to c_2, which is in turn preferred to c_3, then c_1 is preferred to c_3. What we are doing is making an inference according to a principle of logic, namely that implication should be transitive. Given the framework in which such maximization is conducted, this approach provides a strong tool for guiding decision-makers.

Some of the problems with this approach have been discussed already. An important point when comparing it with cost-benefit analyses as a decision aid, is that preferences have to be specified for all consequences, which is a difficult task in practice, and more important, not necessarily something that management would like to do. Refer to the example above where utilities were established for cost and loss of life. Specifying a value of a life is required. This value is related to an arbitrary person in the population, not a specific individual. Note that in a cost-benefit analysis the value of a statistical life is of interest, which is defined here as the expected cost relative to the expected saved lives, which is conceptually not the same as the number in the utility approach. In practice, however, these numbers could be the same. The point is that the utility approach requires values (utilities) to be assessed for all consequences Y, whereas for the cost-benefit approach, the value judgements to be made by the decision-makers relate to $P(Y)$, the probability assignments, and not Y. Thus in the cost-benefit case we assess values in a world constructed by the analysts, not the real world as in the utility-based approach. Usually it is much easier to relate to this constructed world, as we can employ appropriate summarizing performance measures. The simple example in Section 5.1.2 demonstrates this.

A cost-benefit analysis requires us to specify the value of a statistical life, not the value of a life. And that is not the same. We should acknowledge that a life has in principle an infinite value; there should be no amount of money that a person would find sufficient to compensate the loss of his son or daughter, and society (or a company) should not accept a loss of a life with certainty to gain a certain amount of money. On the other hand, a statistical life has a finite value, reflecting that decisions need to be taken that balance benefits and risks for loss of life. The value of a statistical life is a decision-support tool. Now we are to take a decision influencing the future; then by assigning a value to a statistical life, it is possible to obtain an appropriate balance between benefits and risks. When the future arrives, we would focus on the value of life and not the value of a statistical life. For example, if a person becomes ill, the money used to help this person would not be determined by reference to the value of a statistical life, but to the value of the person's life. For this person and his or her closest family it is infinite, but for someone else, it is bounded. What we refer to here is the value of loss that we are willing to accept, given that this benefit is present. What we are willing to pay, to obtain a benefit, is something else. How much should society be willing to pay to save a (statistical) life? In a cost-benefit analysis, focus is usually on this willingness to pay, rather than willingness to accept. But this is not an obvious approach as it means a standpoint with respect to what is the starting point. For example, do the public have a right to a risk-free life, or does industry have a right to cause a certain amount of risk? In the former case, the public should be compensated (using willingness to accept values) by a company wanting to generate risk. In the latter case, the public should compensate (using willingness to pay values) a company keeping its risk level below the maximum limit (Bedford and Cooke 2001: 282).

The use of lotteries to produce the utilities is the adequate tool for performing trade-offs and reflecting risk aversion, but is hard to carry out in practice, in particular when there are many relevant factors, or attributes, measuring the goodness of an alternative. However, tools exist to simplify the assessment of utilities, and one important category is known as multi-attribute utility theory. We refer to Section 5.2.9.

As we discussed in the previous section, we may alternatively perform a multi-attribute analysis without any explicit trade-offs. We assess the various attributes, costs, safety, political aspects, etc., separately and it is a management task to make a decision balancing the costs and benefits. Would that mean lack of coherence in decision-making? Yes, it could in some cases. The ideal is not always attainable. We acknowledge that such a multi-attribute analysis is rather easy to conduct – it works in practice – but the price may be some loss of coherency and traceability in the decision-making process. However, we gain flexibility and in many cases this is of great importance, in particular when the decision situation involves many parties.

There is also multi-attribute analysis with explicit trade-offs that are not based on utilities, see Section 5.2.9.

Bayesian decision theory uses the term 'rationality' in a technical sense, linked to a behaviour satisfying certain preference axioms, including the transitive axiom mentioned above, see Bedford and Cooke (2001); and French and Insua (2000). We use the concept of rationality in a wider sense, in line with Watson and Buede (1987). If we adopt some rules which our statements or actions should conform to, we act in a way that is consistent with them – we act rationally. As there are many ways of defining rules, this means that whether a behaviour is rational will depend on the rules adopted. We find that the rules of Bayesian decision theory constitute a sensible set of rules, but it follows from our definition of rationality that people who do not abide by the percepts of decision theory are irrational; they may have perfectly sensible rules of their own which they are following most rationally. Consequently, if you were to adopt the structure for decision-making presented in this chapter, you would behave rationally, according to the rules set by that structure.

Again we emphasize that we work in a normative setting, saying how people should structure their decisions. We know from research that people are not always rational in the above sense. A decision-maker would in many cases not seek to optimize and maximize his utility, but he would look for a course of action that is satisfactory. This idea, often known as bounded rationality, is just one of many ways to characterize how people make decisions in practice. See the bibliographic notes for some relevant literature.

Despite the fact that managers often behave in conflict with goals, criteria and preferences, we believe that decisions can be improved by a proper structuring of the decision-making process. Our way of thinking provides some guidance on this process, it does not describe a detailed procedure, but balances the need for consistency and flexibility.

A decision, and a decision-making process, may be regarded as good by some parties, and bad by others. Return to the development of an offshore oil and gas field. One particular development concept could be considered good for the oil company, but not so attractive for society as a whole as it could mean a rather high environmental risk and less activity onshore compared to another development alternative. But decisions need to be taken, and proper consideration needs to be given to all relevant parties. Yet such considerations are not easily transformed into a mathematical formula and explicit trade-offs. In many cases, especially when dealing with societal risk problems, we believe that more can be gained by deliberation, where people exchange views, consider evidence, negotiate, and attempt to persuade each other. Deliberation that captures part of the meaning of democracy and contributes to making decisions more legitimate, is also a part of our decision framework, although not explicitly shown in Figure 5.1.

The tools we have discussed for structuring the decision-making process and providing decision support can also be used for decisions made by a group. Individuals still have to decide how they will act, even if the context is organizational politics. Decision analyses, which reflect personal preferences, would give insights to be used as a basis for further discussion within the group. Formulating the problem as a decision problem and applying formal decision analysis as a vehicle for discussion between the interested parties, provides the participants with a clearer understanding of the issues involved and why different members of the group prefer different actions. Instead of trying to establish consensus on the trade-off weights, the decision implications of different weights could be traced through. Usually, then, a shared view emerges of what to do (rather than what the weights ought to be).

There is much more to be said about the decision-making process, but instead of a general discussion we prefer to illustrate our points through some examples.

5.2 SOME EXAMPLES

5.2.1 Accident Risk

We return to the event tree example in Sections 3.3 and 4.3.4. The analysis group concluded that risk-reducing measures should be considered as the calculated risk is rather high. For a ten-year period, a probability of an accident leading to fatalities is computed to be about 8%. Comparing this figure and the FAR value of 55 with risk numbers for similar activities, risk analysis results and historical numbers, the analysis group has a solid basis for its conclusion.

The next step would be to consider possible risk-reducing measures, including the relocation of the control room. This measure would be the best option when it comes to safety for the operators, but it would also be the most expensive one for the company. An analysis of the cost of relocating the control room was then undertaken. This analysis predicted a cost of 0.4 million dollars, and the uncertainties in this prediction were rather small. Thus the cost per expected saved life is $0.4/0.08 = 5$ million dollars. Other risk-reducing measures were also considered, but their effect was not found to be very good relative to the cost of implementing them.

Sensitivity analyses were conducted to see the effects of some of the model assumptions, for example the number of fatalities in each scenario.

Management used the risk analysis as decision support. The safety level for the operators of the control room had been an issue for a long time, and there had been strong pressure from labour organizations to implement some risk-reducing measures. This, together with the clear message from the risk analysis, convinced management that relocation of the control room was required, despite the fact that the cost per statistical saved life was quite high.

No risk acceptance (risk tolerability) criterion was used in this analysis. The principle adopted was that risk should be reduced to a level as low as reasonably practical (ALARP). That means a type of cost-benefit analysis. If an acceptance criterion is defined, risk is considered unacceptable if the calculated risk exceeds a certain level, and risk-reducing measures should be implemented. For most people that are not experts in risk analysis, the use of risk acceptance criteria seems adequate. One specifies certain criteria and draws conclusions based on the calculated risk exceeding these criteria or not. The use of risk acceptance criteria shows commitment – the company would under normal circumstances implement risk-reducing measures if the criteria were not met. If the ALARP principle applies, the lack of absolute criteria could result in inconsistencies and the acceptance of higher risk levels as too high costs is always a convenient argument to use. The problem is that risk acceptance criteria gives a strong form of mechanical thinking when dealing with difficult decision situations involving various aspects of cost and benefit. When decisions are to be taken on the need for risk-reducing measures, it is not sufficient to look just at the calculated risk; other aspects also need to be considered, such as the cost of the measures and the perception of risk. In addition, using risk acceptance criteria could give the wrong focus – the main issue would be to achieve risk acceptance instead of a drive for improvement. If the calculated risk is extremely high – it is considered intolerable – measures would always be implemented, as in the three-region approach of Section 2.1.2, p. 22.

The results of the analysis should be presented in a form that is suitable for the target group. In this case there are two such groups: the company management and the workers being exposed to the risk. The workers also include the labour organization. The presentation so far has been directed at the management; now let us consider the problem of communicating the results to the workers. We cannot expect these people to be familiar with risk analysis.

It is necessary to give the result a form that is easy to understand and which gives confidence and trust. We believe that the following principles should be adopted:

- Focus on observable quantities. Probabilities and expected values should be presented with care.
- Highlight measures that are taken to prevent accidents from occurring and which reduce their consequences if they should occur.
- Use comparisons with familiar activities to illustrate the calculated risk level.

Here is an example presentation based on these principles:

> We cannot ignore the risk of an ignited leakage scenario resulting in fatalities. The company acknowledges this – operating an offshore production installation means some exposure to risk. Substantial work has been done to prevent such a scenario from occurring, including a comprehensive inspection system for pipes and tanks, and a training programme for operation and maintenance personnel.
>
> During a ten-year period it is not likely that such a scenario would occur, but there are uncertainties. And we consider these uncertainties to be so significant that measures need to be implemented. The most effective way turns out to be removing the control room from the process area.
>
> The risk analysis has calculated a probability of about 10% for an ignited leakage scenario resulting in fatalities during a ten-year period. Compared to what is normally considered a reasonable safety level for workers, this is a rather high risk. The cost of removal is about 0.4 million dollars, but the company finds that the cost is not grossly disproportionate relative to the safety improvement obtained. The company will therefore remove the control room from the process area.

If the conclusion had been not to implement risk-reducing measures (given a different risk picture), the arguments would have been similar, but now it would be emphasized that we are confident such a scenario would not occur. Reference could also be made to the cost per statistical saved life, as well as to the measures implemented to avoid the occurrence of the scenario and to other activities where such confidence exists.

Whether that would convince the workers and the labour organizations, is another question. As there is no true risk, the company would need to acknowledge that there could be different views and perceptions. The next example examines this further.

5.2.2 Scrap in Place or Complete Removal of Plant

A chemical process plant is to be decommissioned. The plant is old, and the company that owns the plant would like to scrap and cover the plant in place. People that live close to the plant, environmentalists and some of the political parties are sceptical about this plan. They fear pollution and damage to the environment. Large amounts of chemicals have been used in the plant process. The company therefore looks into other alternatives besides scrapping in place. One alternative is immediately considered to be the most interesting:

> All materials are removed from the plant area and to the extent possible reused, recycled and disposed. A major operation is conducted related to lifting and transport of a huge plant component. The lifting and transport is difficult and there is concern about the operation resulting in a failure with loss of lives and injuries. There are large uncertainties related to the strength of the component materials; if the lifting operation is commenced, it could be stopped at an early stage because it cannot be completed successfully. A considerable cost is associated with this initial phase of the operation. The cost associated with full removal is very large. We refer to this as the removal and disposal alternative.

The company is large and multinational. Due to the tax regime, the state will pay a major part of the removal and disposal costs. Nevertheless, the company makes the final decision on how the plant will be decommissioned. The authorities, through the supervisory bodies, see to that laws and regulations are met. The company seeks a dialogue with these bodies to ensure the parties have a common understanding of the regulations' requirements.

The question is now what principles, what perspective, should be adopted to choose the 'best' alternative, and in particular how risk and uncertainty should be approached. Here are some more specific questions:

- How formalized should the decision-making be?
- Should risk and uncertainty analyses be carried out?
- If such analyses are being undertaken, how should the analysis results be presented, and how should the results be used in the decision-making process?
- Should risk acceptance criteria be defined?
- Should the ALARP principle and cost/benefit analyses be adopted?
- Should one attempt to use utility functions to weight values and preferences?

Furthermore, how should the environmental organizations present their view on risk and uncertainty associated with possible pollution for the scrapping in place alternative? How should the politicians express their view; and the supervisory bodies?

Within the company, a group of competent personnel were asked to provide an advice to the top management on how to approach the problem.

The company decides that its decision is to be based on an overall consideration of technical feasibility, costs, accident risk, environmental aspects, and effects on public opinion. A more formal decision-making process with a one-dimensional cost-benefit parameter was discussed, but it was not considered appropriate as one would expect great differences in value judgements related to the environment, accident risk, etc. A utility-based approach was also considered, but it was soon found to be inadequate. The company would not be willing to use time and resources to establish preferences and utility values over consequences with attributes related to costs, lives, long-term exposures, environmental damage, etc. Any attempt to explicitly compare the possible damages and losses with costs would be extremely difficult to communicate. Risk acceptance criteria were not used, as the situation requires full flexibility with respect to weighting the different costs and benefit dimensions. Before the analyses are conducted, why introduce constraints beyond the legal and regulatory requirements? Studies and evaluations of the different alternatives were carried out addressing aspects such as technical feasibility, costs and safety. The studies were carried out by recognized consultants. Of the results obtained, we briefly look into the cost and accident risk analyses.

> The predicted cost of the scrapping in place alternative is 10 million dollars, with a 90% uncertainty interval given by ±5 million dollars. This means the analyst who has done the assessment is 90% confident that the cost would be within the interval [5,15] million dollars. For the removal and disposal alternative the corresponding numbers are 100 and [50,150], thus substantially larger costs.

> When it comes to accident risk, most concern is related to the removal and disposal alternative. The focus is on successful operation. And if the operation is not successful, what will be the consequences, loss of lives and injuries? Risk analyses have been conducted and they conclude there are large uncertainties related to whether the lifting operation can be executed without losing the component. Unproven techniques have to be used for the operation, and there are large uncertainties in the quality of the component materials. These uncertainties can be reduced by detailed analysis and planned measures. The remaining uncertainties in relation to the event 'the lifting operation is successful' are expressed by a probability of 1/20. When ensuring technical feasibility of industrial projects, an unreliability of 1/20 is considerably higher what is normally accepted. But this is a unique type of operation and it is difficult to make good comparisons. The operation does place personnel at risk, but the risk level is in line with typical values for industrial projects. Transportation of the component is not seen as a safety problem if the planned measures are implemented.

> Following the plans for scrapping the plant, there will be no environmental problem; all chemicals will be removed. Measurements will be carried out to ensure no pollution is present.

> Several environmental organizations and the people that live in the neighbourhood of the plant are sceptical about the company's conclusions on the

environmental impacts. How can one be sure that all chemicals are removed? They refer to the bad reputation this company has from similar activities internationally, and the fact that it could be technically difficult to ensure that no surprises occur in the future if the company implements its plans.

The political parties have different views on this issue. All parties say that the company must remove all chemicals so that people can feel they are safe, but there are different opinions on whether this means the removal and disposal alternative should be chosen.

The company makes an overall evaluation of all inputs, studies and statements from a number of groups, and the dialogue with the supervisory bodies, and concludes that the best alternative is scrapping in place. As there are no safety and environmental problems with this alternative, the additional cost of the removal and disposal alternative cannot be justified. The company is convinced that its procedures for removing all chemicals would work efficiently – measurements will be carried out to ensure there is no pollution – but it respects that others are concerned, especially the people that live close to the plant. The company recognizes the importance of this problem, but cannot see that it justifies the rather extreme cost increases implied by the removal and disposal alternative. If this alternative is chosen, one could use a substantial amount of money (10–30 million dollars) and risk not succeeding at all. The company is not concerned that its reputation will be damaged by the decision on scrapping in place as it has been open about all facts and judgements made.

Whether the chosen alternative would satisfy the requirements set by the authorities would depend on the documentation the company can provide. It turned out in this case that the supervisory bodies required more studies to reduce the uncertainty related to the environmental impacts of scrapping in place. The final outcome would then largely be determined by the supervisory bodies' consideration of this uncertainty, and that consideration could be influenced by environmentalists. Seldom do sharp limits exist that say what is acceptable and what is not, and then the issue and the discussion will give an impression there is significant uncertainty over the environmental impacts.

Given the new documentation, and some additional measures to reduce uncertainty, the supervisory bodies found the chosen alternative to satisfy the requirements set by the authorities.

Not all environmentalist and not all people living close to the facility were happy about this conclusion, but they could not reverse it. They tried physically to stop the operations, but after a short delay, the facility was scrapped and covered in place. So far, no pollution has been notified.

In this case the company, through the consultants, presented risk according to the principles of Chapters 3 and 4. This approach represents a more humble attitude to risk than is often seen in similar situations, as the risk picture established covers predictions and uncertainty judgements. Traditionally, the company would have presented the results from the analysis as representing the truth, the risk associated with the activity, and claimed that laypersons, including the people in the neighborhood of the facility, were influenced by perceptional factors.

It is well known that risk perception and acceptance are influenced (negatively) by factors such as:

- involuntary exposure to risk;
- lack of personal control over outcomes;
- lack of personal experience with the risk (fear of unknown);
- effects of exposure delayed in time;
- large uncertainties related to what will be the consequences;
- genetic effects of exposure (threatens future generations);
- benefits not highly visible.

In our case, many of these factors are relevant, perhaps all. When a person draws conclusions about acceptable safety or risk, he will take into account his own judgement of risk, i.e. his probabilities of observable quantities, the results from risk analyses that provide the analysts' judgements about observable quantities, as well as perceptional factors as listed above. In many cases, including the one presented here, the third point is the most important. If the risk analysis shows a small risk, it does not matter if you feel fear.

It is typical that many experts judge nuclear power as relatively safe, whereas the layperson ranks nuclear power as very dangerous. Who is right? If our starting point is that there exists a true underlying risk, as in the classical framework, we could in theory compare with this risk. The problem is that this risk is unknown and has to be estimated, and the estimate is subjective and very uncertain. Accident statistics give some information about the risk, but here we are dealing with rare events that have not all occurred yet, therefore a risk analysis has to be conducted to estimate risk. The risk analysis is based on experience data and risk analysis methods, but we cannot avoid subjective elements in the analysis process.

The traditional thinking has been that there is a sharp distinction between real risk and risk perception. The company gives the impression that it knows the truth, i.e. the correct risk, and it argues that with increased knowledge and proper communication others would also see the truth. Many see this attitude as provocative, because risk analysis expresses opinions as well as facts, and this is also true for the classical approach to risk analysis.

Adopting our principles of risk analysis, none can say that they have found the true risk numbers, since risk is a judgement about uncertainties. In this way risk analysis is a tool for argument and debate more than a tool for presenting the truth. We have to accept that different persons and parties could have different views. But even if we can agree on the probability assignments, this does not mean that we agree on saying that risk is high or low. Judging the danger as high is a result of finding the occurrence probability of certain events as large relative to the associated consequences. We cannot therefore argue that it is wrong to say the risk associated with nuclear power is considered very high even though the probability that a serious accident will occur is judged very low on an absolute scale. There is a possibility of extreme consequences, and even a small probability may then be sufficient for saying that the risk is high.

The environmentalists, the politicians and the supervisory bodies express their views on risk by discussing uncertainties, and that is consistent with our approach. In quantitative risk analysis this discussion is based on predictions and assigned probabilities, but any judgement of uncertainty is a way of describing risk according to our principles. The weight of such judgements and discussions is strongly affected by the way they are supported by knowledge and facts.

The above example illustrates what is sometimes called an acceptable risk problem. It typically involves experts, public, politicians and other interested parties such as environmentalists. There are several reasons why it is difficult to make decisions in such a context:

- The benefits of the activities could be unclear or disputed or not shared.
- The potential hazards are large and the uncertainties are large.
- The advantages and disadvantages do not fall in the same group or in the same time frame.
- Decisions are seen to be forced upon smaller groups by a higher or faraway authority.
- There is argument between experts and others about hazards and risks.

Extensive political conflict, complexity of a problem and media coverage may strengthen the effects of these factors. Under these circumstances, decisions may not be accepted by society and the position of authorities and the experts who advise them are called into dispute.

Often the experts are seen as acting on behalf and under control of an interested party, producing results and advice that this party wants to hear and see. In some cases the same experts are seen to be in the camp of the other party by all other parties, no matter how objectively they try to establish the facts and formulate their findings.

Risk communication was seen as an instrument to overcome the difference between perceptions of the experts and the public. It was believed that more information and teaching would make society understand. But it is not surprising that society was rarely convinced by this form of communication. What is required is trust and developing confidence in a bidirectional process. If one party tells the other how things are, what the true risks are, they will destroy the trust they are seeking to build.

5.2.3 Production System

The starting point is the production risk example of Section 3.2.2. An oil company evaluates several design options for a gas production system. Let us say that the question is about two or three compressor trains. The production risk analysis produces for each of these alternatives a prediction of the production volume (loss), with associated uncertainties. A histogram representing the uncertainty distribution of the production for each alternative is presented; Figure 5.4 shows a typical example. Other performance measures are also studied. Based on these studies an economic analysis is carried out, including a cash flow

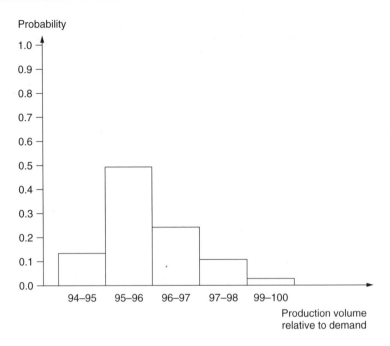

Figure 5.4 Uncertainty distribution of the production volume for a specific alternative

analysis producing expected net present values (NPVs). These analyses provide decision support, they give valuable insight into the uncertainties related to future production volumes.

The analyses are based on a number of assumptions, models, simplifications and judgements. When making its decision, management needs to take these into account. For example, the analysis does not incorporate losses due to loss of reputation by poor performance, nor options for increased production sale. Such factors would be evaluated in parallel to the production risk analysis but not integrated with this analysis.

5.2.4 Reliability Target

Production safety (deliverability) management by using acceptance criteria (targets) has been applied, or at least attempted, in several petroleum development projects. The proposed approaches differ somewhat but the following three approaches are typical of the general philosophy that seems to prevail:

- *Approach 1*: define a system production safety target and allocate subtargets to the items of the system.
- *Approach 2*: define a system production safety target and use system design optimization to obtain an acceptable solution.
- *Approach 3*: define a feasible concept of the system, calculate its production safety and call it the target.

All these management approaches have fundamental shortcomings when it comes to solving the actual problem at hand. These shortcomings will be demonstrated by a specific example and a discussion of the general nature of planning complex production systems.

The purpose of the project in this example was to produce natural gas from an offshore gas field and bring it ashore for delivery to purchasers. After some initial studies had been carried out, a decision was made to develop the field with a production platform and subsea pipelines for gas transportation. It was further agreed that a systematic treatment of production safety (risk) would be of benefit to the project. A study was therefore commissioned, with a conceptual description of a proposed design as input, to define a production safety acceptance criterion (target) for the platform.

The first obstacle the study team ran into was related to the definition of the overall system boundary. Their task was to define a target for the platform; however, it turned out that the performance of the production system as experienced by the gas purchasers would be very different from the performance of the platform when viewed in isolation. This was due to the large internal volume of the pipeline transportation system and the compressibility of gas, which enabled the system to be used as a buffer storage. The inherent overcapacity of the overall system thereby enabled production outages below certain volumes to be recovered by the system at the point of delivery. Hence the study team recognized that the transportation system had to be considered in the definition of an overall production safety target.

But what should be the target for the production safety of the total system? This question resulted in considerable discussion, because no one was immediately able to assess the consequences of choosing a specific figure. Nor was it possible to determine a corresponding requirement for the platform's production safety.

The following main conclusion was accordingly drawn by the study team:

> It would be impossible to know which level of production safety should be preferred as a target without knowing what it would require and what it would yield in return to achieve all the possible levels.

As a result, the possibility of using approach 1 and approach 2 was abandoned by the study team. An attempt was subsequently made with approach 3, but it was soon realized that it would lead to little more than an adoption of a coarsely sketched concept as an optimal solution. Any subsequent action to optimize the design would require the targets to be changed, and a moving target would lose its intended meaning. Consequently, the whole concept of production targets was abandoned for this project.

Shortcomings of the production safety target approach

We conclude that any attempt to apply a production safety target approach to the problem of planning a complex oil/gas production system is a failure to recognize the primary objectives of the activity as well as the basic properties of the planning problem itself. To substantiate this statement, let us first consider what the objectives of a project might be.

To begin with, we should acknowledge that the categories 'correct' and 'false' do not apply to a given design of a production system; we can only say that it is a good or a bad solution and this to varying degrees and maybe in different ways for different people. Likewise, there will be different conceptions with regard to the objectives of an enterprise, but the bottom line of any oil and gas project is still profit. Profit is the main objective and driving force of the industry. Other conditions, such as production safety, may have to be fulfilled to some extent, but these are only a means of reaching the primary goal. From this line of argument, we can conclude that production safety should not be treated as an objective in its own right.

One might still ask whether production safety targets could not be used merely as a guideline to attain the objectives. It is often said about stated production safety targets that they are not intended as absolute levels, but only as a means of communicating a certain policy. Unfortunately, the relationship between the production safety target and the policy is seldom very well defined. Furthermore, a lot of good managers and engineers have a tendency to interpret a figure called a target as something one is supposed to attain. Specifying an absolute level without really meaning it could therefore prove a dangerous practice. It may restrain innovation and sound judgement, and result in an unnecessarily expensive design. But what about giving a production safety target as a range or a distribution? Or why not go all the way and use a qualitative statement only: Our target is to achieve normal production safety. As we can see, the whole thing is starting to get rather vague. Consequently, a production safety figure is not suitable as a policy guideline.

We conclude that, as a general rule, production safety targets should not be used at all. Instead a more cost-effective approach should be adopted, where attention is focused on finding the most economic (profitable) solution, rather than on attaining unfounded targets.

5.2.5 Health Risk

In this section we study a decision problem related to the health risk example studied in Section 4.3.4. We test a patient when there are indications that he has a certain disease. Let X be 1 or 0 according to whether the test gives positive or negative response. Furthermore, let θ be the true condition of the patient, the state of nature, which is defined as 2 if the patient is seriously ill, 1 if the patient is moderately ill, and 0 if the patient is not ill at all.

Now let us go one step further and follow our patient after both tests have shown a positive response. The doctor then needs to make a decision based on the updated probabilities. We have $P(\theta = 2|X_1 = 1, X_2 = 1) = 0.27$. Similarly, we find that

$$P(\theta=1|X_1=1, X_2=1) = \frac{0.60 \times 0.36}{0.90 \times 0.11 + 0.60 \times 0.36 + 0.10 \times 0.53} = 0.59,$$

HOW TO USE RISK ANALYSIS TO SUPPORT DECISION-MAKING

Table 5.1 Expected portion of normal life expectancy

Decision	Health state	
	$\theta = 2$	$\theta = 1$
d_1	10%	80%
d_2	50%	50%

$$P(\theta=0|X_1=1, X_2=1) = \frac{0.10 \times 0.53}{0.90 \times 0.11 + 0.60 \times 0.36 + 0.10 \times 0.53} = 0.14.$$

We have established the posterior distribution of θ. Thus the highest probability is related to the patient being moderately ill, but there is quite a large probability of the patient being seriously ill, which means he has have to be immediately sent to hospital. If the patient is seriously ill, immediate treatment is necessary to avoid disablement or death. The doctor is facing a decision problem under uncertainty. Should the patient be hospitalized or not? Well, this is a rather simple decision problem, clearly the patient should be immediately hospitalized as the probability that he is ill is so large and the possible consequences so severe if he is not treated. We need no further optimization and evaluation. The risk or uncertainty picture provides a clear message about what to do.

Now, suppose further analysis and testing of this patient at the hospital gives updated posterior probabilities of $\theta = 2, 1$ and 0 equal to 0.3, 0.7 and 0.0, respectively. Two possible medical treatments are considered: d_1, which would be favourable if $\theta = 1$, and d_2, which would be favourable if $\theta = 2$. The expected portion of normal life expectancy given θ and d_i is shown in Table 5.1. We see that if $\theta = 2$, then treatment d_1 would give a life expectancy of 10% relative to normal life expectancy. From these expectations, a utility function can be established, reflecting the preferences of the patient, or alternatively the physician.

Let us look at how we can elicit the utility function for the patient. The starting point for establishing the utility values is 0 and 1, corresponding to immediate death and normal life, respectively. We then ask the patient to compare an expected life length without operation of say 15 years with a thought-constructed operation having a mortality of $x\%$; however, if the operation were successful, the patient would enjoy a normal life with expectancy of 30 years. This exercise is not directly linked to the medical treatment the patient is going to undertake. The patient is asked what is the minimum probability of success from the operation needed to undergo the operation. Say it is 90%. Then this number is the utility value related to a proportion of life expectancy of 50%. Obviously, this probability would be higher than 50% as the patient is 'guaranteed' 15 years of life with no operation, whereas the operation could lead to death. Other utility values are established in a similar way and we arrive at the utility function shown in Table 5.2. Using the expected life expectancy to establish the utility

Table 5.2 Utility function for the two decision alternatives

Decision	Health state	
	$\theta = 2$	$\theta = 1$
d_1	0.40	0.95
d_2	0.90	0.90

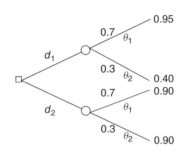

Figure 5.5 Decision tree for the decision problem summarized in Table 5.2

function is a simplified approach as it ignores the quality of life, for example. But it is not viewed as too gross a simplification.

Figure 5.5 shows a decision tree for the decision problem. The tree grows horizontally from left to right. Beginning from the left, there are two decision alternatives represented by two branches of the tree. This describes the decision structure, but at the end of the present terminal branches (each of which is either d_1 or d_2) we add two others, labelled θ_1 and θ_2, corresponding to the health state of the patient. The points where the branches split into other branches are called decision nodes or random nodes, depending on whether the branches refer to possible decision choice alternatives or uncertainties (of observable quantities). These two types of nodes are represented by a square and a circle, respectively. According to the utility paradigm, the decision maximizing expected utility should be chosen.

We find that the expected utilities for the two decisions, u_1 and u_2, are given by

$$Eu_1 = 0.40 \times 0.3 + 0.95 \times 0.7 = 0.785,$$

$$Eu_2 = 0.90 \times 0.3 + 0.90 \times 0.7 = 0.900.$$

Thus decision d_2 should be chosen. Of course, this is what the mathematics says. The analysis is based on simplifications of the real world, and it is based on the preferences of the patient only. The costs involved are not reflected. The physician must also take this into account, if relevant, when he establishes his utility function. In most cases the patients and physicians agree on which treatment to undertake, but conflicts could occur. Utilities are a tool for communicating values, but they do not solve the difficult problem of dealing with different preferences between patients and physicians.

It is possible to use the same type of reasoning when dealing with reduced quality of life. The person specifying the utility function then needs to compare a number of years of reduced quality of life with a normal life.

A utility-based approach ensures coherency in medical decision-making. Viewing the total of activities in medicine, there is a strong need for using resources effectively and obtaining optimal results. The introduction of utility functions could be somewhat standardized to ease the assessment, as a number of examples can be generated.

Could we not have used a multi-attribute analysis or a type of cost-benefit analysis instead? Why not produce predictions and assessed uncertainties related to the result of the treatment, costs, etc.? Is it really necessary to specify a utility function? The figures in Table 5.1 are very informative as such. They provide valuable insights and a good decision basis. If we chose not to introduce a utility function, we would evaluate the predictions and assessed uncertainties, but we would give no numerical utility value on the possible outcomes. This could be satisfactory for patients as they do not need to think about coherency. For the hospital and the society, however, coherency is an issue.

5.2.6 Warranties

We consider the exchange of items from a large collection of similar items N between a manufacturer (seller) and a consumer (buyer). A warranty contract pertaining to the item reliability is sought. The following is a typical warranty contract in many transactions.

Let n be the number of items that the buyer (B) would like to purchase. These items are supposed to be identical. Each item is required to last for τ units of time. We suppose that the buyer is willing to pay x dollars per item, and is prepared to tolerate at most z failures in the time interval $[0, \tau]$. For each failure in excess of z, the buyer B needs to be compensated at the rate of y dollars per item. In effect, the quantity τ can be viewed as the duration of a warranty.

Below we sketch how the seller A can proceed to specify initial values of z and y.

Suppose that it costs c dollars to produce a single unit of the item sold. Then if the buyer B experiences z or fewer failures in $[0, \tau]$, A's profit would be $n(x-c)$. However, if B experiences i failures in $[0, \tau]$ with $i > z$, then A's liability will be $(i - z)y$. Let p be the proportion of failed units in the large population of items. We refer to p as a chance – it is an observable quantity. Then if $P(i)$ denotes the chance of exactly i failures in the time interval $[0, \tau]$, we have

$$P(i) = \binom{n}{i} p^i (1-p)^{n-i},$$

i.e. the number of failures is binomially distributed with parameters n and p, were we to know p. Furthermore, A's expected liability is

$$\sum_{i=z+1}^{n} y P(i).$$

From these formulas the expected profit, given p, would be

$$n(x - c) - \sum_{i=z+1}^{n} y P(i).$$

If the seller has strong background information concerning the failure frequency of the items, p could be considered known. If that were not the case, the seller would need to assign an uncertainty distribution (a prior distribution) H on p, and compute the unconditional expected value:

$$\int \left[n(x - c) - \sum_{i=z+1}^{n} y P(i) \right] \mathrm{d}H(p).$$

This analysis can be used by the seller as a basis for identifying values of z and y that they would find acceptable. A similar analysis can be carried out for the buyer, and we can discuss what should be a fair contract; see Singpurwalla (2000).

5.2.7 Offshore Development Project

Let us reconsider the decision problem discussed in Chapter 1, where two concepts, A and B, for the development of an oil and gas field are assessed. To provide a basis for choosing an alternative, a multi-attribute analysis is carried out based on separate assessments of relevant factors such as technology development, market deliveries and regularity, investment costs, operational costs and safety and environmental issues. Let us look at some of the assessments without going heavily into the details.

Technological development

This expresses the value created by the alternative with regard to meeting future technology needs for the company. Alternative A is risk-exposed in connection with subsea welding at deep water depth. A welding system has to be developed to meet a requirement of approximately 100% robotic functionality as the welding must be performed using unmanned operations. The alternative is risk-exposed, meaning that the welding system development could cause delay and consequently increased costs, and it could be more costly than expected. But the risk exposure is considered moderate, as there is a fallback based on manned operations as an emergency option. This will prevent major schedule effects on the production start date.

Schedules

The schedule for offshore tow-out is tighter for alternative B than for alternative A. For alternative B a probability of 0.15 is assigned for a delay in production start. The assigned probability distribution for the number of days delay is 0.10, 0.05 and 0.01 associated with delay periods of 15 days, 45 days and 75 days, respectively. Assuming a cost of 50 million dollars per month delay, an expected loss of 7.5 million dollars is computed. The costs associated with delay for alternative A are considered negligible compared to B.

Regularity to market

The gas regularity requirement set by the market is 99%. The predicted (expected) market deliveries are about 99.0% for alternative B and 99.5% for alternative A, but there are significant uncertainties involved. These uncertainties are expressed by probability distributions of the market deliveries, similar to Figure 5.4. If the deliveries cannot be met, other sources will be used, including gas from an onshore gas storage. The expected yearly costs of the back-up gas are calculated and these costs are transformed to expected NPV values.

Investment

The expected investment costs for the two alternatives are found to be about the same, 3 billion dollars. Uncertainties in the investment costs are presented using simple histograms, analogous to Figure 5.4. An advantage of alternative A is more time for plant and layout optimization. The main potential includes reduction in management and engineering man-hours, reduction in fabrication costs, and optimization of the process plant (arrangement of plant, use of compact technology, number and size of compressors and generator drivers). The difference in expected upside potential between the two alternatives is 0.4 billion dollars, in favour of alternative A.

Operating and maintenance costs

The expected operating and maintenance costs are approximately the same for the two alternatives, but the uncertainties are larger for B than for A as there is less experience of using concept B. The uncertainties are quantified by probability distributions similar to Figure 5.4. For both alternatives there are some upside potentials and downside risks. These are presented as expected reduced costs and expected increased costs.

Reservoir recovery

There is no major difference between the alternatives on reservoir recovery.

Environmental aspects

For each alternative, predictions are presented showing emissions (in tonnes per year) to air from turbines, diesel engines, flare, and loading of oil and condensate. Alternative B has the greatest potential for improvement with respect to environmental gain. New technology is under development in order to reduce emissions during loading and offloading. Further, the emissions from power generation can be reduced by optimization. Otherwise, the two concepts are quite similar with respect to environmental aspects.

Safety aspects

For both alternatives there are accident risks associated with the activity. The analysis shows a slightly higher accident risk for alternative A than for alternative B. Both alternatives would be able to comply with the overall safety requirements. Risk-reducing measures need to be identified, evaluated and implemented on the basis of evaluations of cost and benefit.

External factors

Concept A is considered to be somewhat advantageous relative to concept B as regards employment, as a large part of the deliveries will be made by the national industry.

Summary of the analyses

The NPV values for schedules, regularity, investment, and operating and maintenance costs are presented in a table and integrated. In addition the various other factors are given a + or − depending on which alternative is found to be favourable, and together this provides a summary of the analyses and a basis for making a decision on which alternative to choose.

Many details have been omitted from this analysis report, but it does give an impression of the main line of thinking.

5.2.8 Risk Assessment: National Sector

The task is to develop an approach, a methodology, for assessing the safety level and to identify trends in a national branch or sector, for example an industry. The purpose of the methodology is to improve safety by creating a common understanding and appreciation of the safety level and thus provide a basis for decision-making on risk-reducing measures. The aim is to build consensus through assessments, participation and commitment. Furthermore, by having an increased focus on occurrences that may result in accidents, it is hoped that the number of such occurrences will be reduced.

Now, how should we do this assessment? We restrict attention to large-scale accidents leading to fatalities.

We interpret the safety level as uncertainties about the world and the occurrence of accidents and losses. To assess these uncertainties, some basic principles need to be established. The starting point for the assessment should be the measurement of some historical accidents. As far as possible, these data should be objective data. Secondly, we need evaluations, based on these data and other sources. We acknowledge that assessing the safety level cannot be based on hard data only. Safety is more than observations. We need to see behind the data and incorporate aspects related to risk perception.

HOW TO USE RISK ANALYSIS TO SUPPORT DECISION-MAKING

There are three basic categories of data that can be used:

- loss data, in this case expressed by the number of fatalities;
- risk indicators (hazards) such as major leaks and fires;
- risk indicators on a more detailed level, reflecting technical, organizational and operational factors leading to hazards.

We should collect and analyse data from all three categories. They provide different types of information. Each shows just one aspect of the total safety picture, and if viewed in isolation, data from one category could give a rather unbalanced view of the safety level. We face uncertainties related to a vast number of large-scale accident scenarios, but fortunately we have not observed many of these accidents. Using the historical, observed losses as a basis for the uncertainty assessments could therefore produce rather misleading results. On the other hand, using the risk indicators on a detailed level, as a basis, would also be difficult as they could be of poor quality. Do the indicators reflect what we would like to address? Is an increased number of observations a result of the collection regime or the underlying changes in technical, organizational and operational factors? We regard measurements of the hazards as providing the most informative source for assessing the safety level. There is not a serious measurement problem and the number of observations is sufficiently large to merit an analysis.

Let x_{ij} be the number of hazards observed of type i in year j, $i = 1, 2, \ldots, k$, $j = 1, 2, \ldots, r$. As an example suppose that the data for $i = 1$ is given by 6, 9, 9, 12, 13. To analyse these data, we should adopt the ideas outlined in Section 2.1. Here are the main points:

- Any observed trend in the number of hazards, such as in the example above, should be examined to identify what caused this trend.
- As a screening method for use of resources, a procedure should be defined to identify the hazards having strong trends.

A simple procedure is based on the use of a Poisson distribution to assess uncertainties. Again consider the numerical example for $i = 1$. Suppose we have the observations 6, 9, 9 and 12, and we would like to assess the uncertainty related to the number of hazards for the coming year based on these observations. For simplicity, suppose that the number of exposed hours does not vary from year to year. Then, following an argument as in Section 4.3.4, we use a Poisson distribution with mean 9 and obtain a 90% prediction interval [5, 14]. As the year 5 observation is included in this interval there is no alarm. But an alarm would be given when assessing the uncertainties of year 4 and 5 based on the three previous years. The prediction interval is [14, 24] and the observation is $12 + 13 = 25$.

An integrated yearly risk indicator R can be developed based on the data x_{ij}. It is given by the formula

$$R_j = \sum_{i=1}^{k} v_j x_{ij},$$

where v_j is a weight, reflecting the expected number of fatalities given the occurrence of the hazard j. This expectation is derived from risk analyses.

A group of recognized people with strong competence in the field of risk and safety, are established to evaluate the data observed. These data include the event data and indicators mentioned above as well as other data, reflecting for example the performance of the safety barriers and the emergency preparedness systems. Also attention is given to safety management reviews and results from analysis of people's risk perception. Based on all this input, the group draws conclusions about the safety level, status and trends.

In addition, a group of representatives from the various interested parties discuss and review important safety issues, supporting documentation and views of the status and trends in general, as well as the conclusions and findings of the expert group. The combined message from these two groups provides a representative view on the safety level for the total activity considered. And if consensus can be achieved, this message becomes very strong.

5.2.9 Multi-Attribute Utility Example

We return to the event tree example in Sections 3.3 and 4.3.4. In Section 5.2.1 we used a cost-benefit (cost-effectiveness) analysis to support decision-making. This is our recommended approach for this example. But other tools are also applicable, for example multi-attribute utility theory. In this section we will show how we can use this theory for the event tree example. The decision alternatives considered are relocation of the control room and not relocation. At the end of the section we compare this tool with the cost-benefit analysis of Section 5.2.1.

First we have to identify the relevant objectives. In this case we can summarize the objectives in two main categories:

- minimize costs (maximise profit);
- avoid accidents.

These objectives can be further divided into categories, giving a hierarchy of objectives. For example, 'avoid accidents' could be replaced by 'avoid fatalities' and 'avoid injuries'. Of course, this is a simplification as accidents are more than fatalities and injuries. We base our analysis on minimize costs and avoid fatalities. These objectives are measured on the attribute scales money (x_1) and number of fatalities (x_2), respectively. The challenge is now to elicit a utility function $u(x_1, x_2)$. Suppose we have established individual utility functions $u(x_1)$ and $u(x_2)$ for the attributes x_1 and x_2. The natural candidate for $u(x_1, x_2)$ is to use a weighted average of these different utility functions, i.e.

$$u(x_1, x_2) = k_1 u(x_1) + k_2 u(x_2),$$

where the weights are k_1 and k_2. The sum of these weights is 1. We will discuss the suitability of this additive utility function later; now we will look at how to proceed when this form is being used.

Let us start by establishing the utility function for attribute x_2, the number of fatalities. Following the procedure illustrated in Section 5.1.2, we give the best consequence, i.e. zero fatalities, a utility value of 1, and the worst consequence, i.e. two fatalities, a utility value of 0. It remains to specify a utility value for one fatality. Using the lottery approach explained in Section 5.1.2, we arrive at a value 0.4, say. We find that one fatality is worse than a lottery having a 50% chance of zero fatalities and a 50% chance of two fatalities. The point is that going from zero to one fatality is worse than going from one to two fatalities.

Next we establish the utility function for the costs. The best consequence is a cost zero, so we give this cost a utility value of 1. The worst consequence we define as 10 million dollars, which is given a utility value of 0. Between these values we use a linear function, as the company's attitude to costs and risk within the interval [0, 10] are expressed by the expected value. Thus the cost 0.4 million dollars of removing the control room has a utility value of $u(0.4) = (10 - 0.4)/10 = 0.96$.

Finally, we need to specify the constants k_1 and k_2. Suppose we think that the values placed on the lives of the two control room operators should be 2 million dollars. Then $k_1 = 5/6$ and $k_2 = 1/6$, as $u(10, 0) = k_2$ and $u(0, 2) = k_1$, and thus k_2/k_1 should be equal to 2/10. Then we can compute the expected utility value for the two alternatives, the relocation alternative and the not relocation alternative. We denote these expectations $E_R u$ and $E_N u$, respectively. From Section 3.3 the distribution of the number of fatalities, Y, related to a one-year period, is given by $P(Y = 2) = 0.0016$, $P(Y = 1) = 0.0064$ and $P(Y = 0) = 0.992$. We consider a ten-year period, which gives the approximate probabilities $P(Y = 2) = 0.016$, $P(Y = 1) = 0.064$ and $P(Y = 0) = 0.920$.

We find that

$$E_R u = k_1 \times u(0.4) + k_2 \times 1$$
$$= 0.833 \times 0.96 + 0.167 \times 1$$
$$= 0.967,$$

whereas for the not relocation alternative we find

$$E_N u = k_1 \times 1 + k_2 \{u(2)P(Y = 2) + u(1)P(Y = 1) + u(0)P(Y = 0)\}$$
$$= 0.833 + 0.167(0 \times 0.016 + 0.4 \times 0.064 + 1 \times 0.920)$$
$$= 0.991.$$

Thus the not relocation alternative has the highest expected value and is therefore the favourable alternative in this decision analysis context. To change this conclusion, the value of the two control room operators must exceed about 7 million dollars, corresponding to $k_1 = 0.6$ and $k_2 = 0.4$.

This analysis is based on costs and the number of fatalities only. Other factors (objectives, attributes) would also be considered, such as risk perception. Following multi-attribute utility theory, or utility theory in general, we should include all such factors – the set of objectives should be complete. Our thinking is, however, more pragmatic as decision analysis is just a tool for aiding decision-making. We acknowledge that other factors would be taken into account, but we find it difficult and inadequate to incorporate them as attributes in the analysis. Restricting attention to costs and number of fatalities, the analysis is just slightly more complex than the cost-benefit analysis of Section 5.2.1. We remember that the cost per expected saved life was equal to 5 million dollars. The utility approach is more complex in the way that it needs the establishment of the utility function, which means stronger management involvement. To some extent, it is possible to standardize the utility functions, thus reducing the work to be done in specific cases. The cost-benefit analysis is based on a predefined performance measure; cost per expected saved life, which is rather straightforward to calculate.

In this example the cost-benefit analysis and the utility approach give basically the same message; if the cost of a (statistical) life is of order 3–5 million dollars, the removal of the control room is favourable. This conclusion is based on an analysis of cost and number of fatalities only. Management performs a review and judgement of the analysis and other relevant factors, then makes a decision.

Now we return to the problem of specifying the utility function $u(x_1, x_2)$. Above we used an additive form $u(x_1, x_2) = k_1 u(x_1) + k_2 u(x_2)$. This form simplifies the analysis, but the question is whether it can be justified. The additive form means that our attitude to risk on each of the attributes does not depend on the other attribute. In this case it is a reasonable approximation as the burdens associated with loss of life should not be influenced by the cost of relocation.

In practice it is often difficult to assess a utility function over several attributes, so a number of alternative approaches have been established to perform the trade-offs. We have already looked at the cost-benefit analyses. Another category of approaches are related to the use of value functions, using some form of preferential independence (Bedford and Cooke 2001: 271). It is common to start by specifying a value function, for example by using a multi-attribute value function of the form[1]

$$v(x_1, x_2, \ldots, x_r) = \sum_{i=1}^{r} w_i x_i,$$

where w_i is a weighting factor of the ith attribute. The w_i encode the trade-offs that the decision-maker is prepared to make between the attributes. Special techniques are developed to determine the weights w_i; see Keeney and Raiffa (1976), Keeney (1991) and French et al. (2001). The value function is then transformed to a utility function, u, for example by the exponential transform of v:

$$u(x_1, x_2, \ldots, x_r) = 1 - \exp\{-v(x_1, x_2, \ldots, x_r)\}$$

$$= 1 - \exp\left\{-\sum_{i=1}^{r} w_i x_i / \rho\right\},$$

where the parameter ρ directly encodes risk aversion. Such an approach would simplify the specification of the utility function, but care is needed to avoid arbitrariness in the specification of the utility function.

In addition we like to mention the analytical hierarchy process (AHP), which is a common approach among practitioners; see Bedford and Cooke (2001: 271) and Saaty and Vargas (2001). The AHP does not have the same strong foundation as the utility-based approach, but it is quite simple to use in practice.

5.3 RISK PROBLEM CLASSIFICATION SCHEMES

Sections 5.1 and 5.2 discussed a number of decision situations where risk and uncertainty need to be addressed. Now we will look at some structures, or classification schemes, for these decision situations that are consistent with our predictive approach. Based on these classification schemes, we will discuss the use of risk and uncertainty analyses, formal decision analyses, and risk management policies.

Section 5.3.1 presents a classification scheme based on the two main factors: potential consequences (outcomes, losses, damages) and uncertainties about the consequences. Section 5.3.2 examines a classification specifically directed at accident risk with the dimensions closeness to hazard and level of authority.

The classification systems provide a knowledge base for structuring risk problems, risk policies and class-specific management strategies. Three major management categories have been applied: risk-based, precautionary and discursive strategies. The risk-based policy means treatment of risk – avoidance, reduction, transfer and retention – using risk and decision analyses. The precautionary strategy means a policy of containment, constant monitoring, continuous research and the development of substitutes. Increasing resilience, i.e. resistance and robustness to surprises, is covered by the risk-based strategy and the precautionary strategy. The discursive strategy means measures to build confidence and trustworthiness, through reduction of uncertainties, clarifications of facts, involvement of affected people, deliberation and accountability. In most cases the appropriate strategy is a mixture of these three strategies.

5.3.1 A Scheme Based on Potential Consequences and Uncertainties

This classification scheme is based on two main factors: potential consequences (outcomes, losses, damages) and our uncertainties about the consequences; in other words, the key factors related to our qualitative, broad definition of risk. From these two factors we establish the seven categories in Table 5.3. These seven categories show a tendency of increased risk, level of authority involved, stakeholder implications, and treatment of societal values. The arrows should be read as tendencies, not as strictly increasing values.

To further characterize the consequence potential, beyond straightforward summarizing measures related to losses and damages (such as economic loss and number of fatalities), we relate it to these factors:

Table 5.3 Risk context classification scheme: read the arrows as tendencies, not as strictly increasing values; S = small, M = moderate, L = large

	Category		Level of risk	Level of authority involved	Stakeholder implications	Treatment of societal values
	Potential consequences	Uncertainties of consequences				
1	S	S/M/L	Low	Low	Low	Low
2	M	S				
3	M	M				
4	M	L	↓	↓	↓	↓
5	L	S				
6	L	M				
7	L	L	High	High	High	High

- *Ubiquity* is the geographic dispersion of potential damages.
- *Persistency* is the temporal extension of potential damage.
- *Reversibility* is the possibility of restoring the situation to the state before the damage occurred.
- *Delay effect* characterizes a long time of latency between the initial event and the actual impact of damage.
- *Potential of mobilization* means violation of individual, social or cultural interests and values generating social conflicts and psychological reactions by individuals and groups who feel afflicted by the risk consequences.

And to further characterize the uncertainties we relate them to these factors:

- the degree of predictability of consequences;
- the difficulty in establishing appropriate (representative) performance measures (observable quantities on a high system level);
- persons or groups that assess or perceive the uncertainties.

Depending on how the problem relates to these factors, different risk policies and management strategies would be required. Thus there is more than one risk policy and more than one management strategy associated with each of the seven categories. However, for some of the categories, there is a typical candidate.

Now we describe and discuss the categories of this classification scheme using the headings in Table 5.3.

(1) Small + small/moderate/large

This category is characterized by situations where the potential for loss or damage is small and the uncertainties related to the consequences are small, moderate or large. Examples are driving a car and work activities at a job. There is typically an established practice for the activities. Note that the term 'small' is a relative concept – an injury or a fatality is not a small consequence as such. In these situations we would pay attention to risks and uncertainties, perhaps perform some simple qualitative risk analyses, buy a safe car to increase robustness in the case of an accident, and look for substitutes. But a formal risk management system for the specific situation, the driver or worker, would in most cases not be introduced. If we consider a large population of such cases, for example the car traffic, a risk management system would be required, but that would be a problem within another category.

(2) Moderate + small

This category is characterized by a moderately large potential for loss or damage and small uncertainties related to the consequences. An example is an investment project for a production system where the future income is strongly influenced by production sales contracts. Risk, uncertainty and decision analyses could be used as part of a risk management system that operates within the framework set by the contracts. Other examples are the anthropogenic effect of climate change and the loss of biological diversity. The risks may not be taken seriously because of the long delay between the initial event and the damage impact. This category needs strategies to build awareness or initiate efforts by institutions to take responsibility. A continuous reduction of risk potential is necessary by introducing substitutes. Risk potentials that cannot be substituted should at least be contained by setting quantities and limitations of exposure.

(3) Moderate + moderate

This category is characterized by a moderately large potential for loss or damage and moderately large uncertainties related to the consequences. Many technological risks belong to this category, such as chemical process facilities. The examples in Section 5.2.3, 5.2.4, 5.2.6, 5.2.7 and 5.2.8 may all be viewed as special cases of this category. The consequences are classified as moderate, not large, as they are bounded, with rather low scores on ubiquity, persistency, etc. The maximum loss or damage can be determined. Uncertainties are considered moderate by risk analysts and others as the phenomena leading to the consequences are largely understood. Risk, uncertainty and decision analyses are used as part of a risk management system. Another example is electromagnetic fields generated by the high-voltage overhead power lines, as judged by many laypersons. Although experts are confident that the possible consequences are small, and thus classify the situation into category 1, laypersons may judge the uncertainties to be rather high. The main principle of risk management in this case should be discursive, which means placing emphasis on strategies to build confidence, reduction of uncertainties, and clarifications of facts.

(4) Moderate + large

This category is characterized by a moderately large potential for loss or damage and large uncertainties related to the consequences. An example belonging to this category is a process plant based on a new type of technology. The example discussed in Section 5.2.2 belongs to this category. Uncertainties are considered large as the phenomena leading to the consequences are not well understood. Risk, uncertainty and decision analyses are used as part of a risk management system. Key elements of such a system would be to improve knowledge, to prevent surprises and to plan for emergency management. Compared to category 3, this category has a stronger element of precaution as the uncertainties are larger.

(5) Large + small

This category is characterized by a large potential for loss or damage and small uncertainties related to the consequences. Smoking belongs to this class. The consequences for society are large, whereas our uncertainties related to possible consequences of smoking are rather small. Risk, uncertainty and decision analyses are used as part of a risk management system.

(6) Large + moderate

This category is characterized by a large potential for loss or damage and moderately large uncertainties related to the consequences. The consequences are large, meaning that the losses and damages are difficult to bound, and high scores are given to one or more of ubiquity, persistency, etc. An example belonging to this category is nuclear energy. Uncertainties are considered moderate by risk analysts as the phenomena leading to the consequences are largely understood. Risk, uncertainty and decision analyses are used as part of a risk management system. A layperson's perception of uncertainty may be in conflict with the experts, and they may classify this situation as belonging to category 7. As the consequences are large, a precautionary principle should be implemented, addressing policies on containment, monitoring, research and development of substitutes. The discursive strategy is also important, to build confidence and reduce uncertainties.

(7) Large + large

This category is characterized by a large potential for loss or damage and large uncertainties related to the consequences. Examples belonging to this category are the greenhouse effect, human intervention in ecosystems, technical invention in biotechnology, and persistent ecosphere pollutants. The consequences are large, meaning that it is difficult to bound the losses and damages, and high scores are given to one or more of ubiquity, persistency, etc. Uncertainties are considered large as the phenomena leading to the consequences are not well understood. It is difficult to establish appropriate performance measures (observable quantities on a high level) describing the possible consequences. Some researchers would refer to the uncertainties as unknown uncertainties or ignorance. Risk and uncertainty analyses can be used to study aspects related to specific performance measures. Decision analyses are not seen as an adequate tool. A precautionary principle should be implemented, addressing policies on containment, monitoring, research and development of substitutes. Key elements of the risk management system would be to improve knowledge and emergency management.

How to use this classification system

Risk and uncertainty analyses, and multi-attribute analyses (with no explicit trade-offs between attributes), are conducted for all categories except category 1. Formal decision analyses are restricted to categories 2 to 6, when found appropriate.

This classification system provides a structure for categorizing situations or problems according to potential consequences and uncertainty. These dimensions characterize the situation or problem to some extent, but the definition of a policy and a management strategy needs to take account of other factors, as discussed above. This is an essential point. Risk management is more than expert assessments of uncertainty and risks. We cannot base our decisions on the results of risk and decision analyses alone. In practice we need to find a proper balance between risk-based strategies, and precaution and discursive strategies.

The above classification structure, with adjusted characterizations of potential consequences and uncertainty, can also be applied in a project risk context to identify a list of critical activities and issues that need to be followed up during the project. The scheme then becomes a tool in the uncertainty management of the project.

5.3.2 A Scheme Based on Closeness to Hazard and Level of Authority

Many actors inside and outside an organization are in one way or another involved in dealing with risks. Decisions involving uncertainty and risk are made at different organizational levels and in a number of settings. Process plant managers encounter situations which force them to make decisions that will seriously affect production goals and accident risk in a conflicting manner. To make satisfactory decisions, they are dependent on decisions by senior management, e.g. in the form of policy statements, about priorities of accident risk versus production goals. Regulatory agencies can be seen to make decisions when imposing new requirements, e.g. to perform risk analysis and deal with risk in specified ways. It is obvious that the context and nature of these decision processes varies significantly. Often decision-makers are constrained in a way that does not allow them to assess risk in detail.

The time and resources available for the decision normally restrict the degree of modelling and refinement in the analysis. Even more important, formal risk analysis is associated with procedures and a work environment setting, which do not conform to all kinds of decision settings. It is obvious that senior managers, with a high and diversified workload, in many cases may not be able to perform structured risk analyses over environmental releases for a number of decision alternatives. The same can be said for flight line engineers encountering shaft wear, with a half-hour time window to complete their inspections and maintenance. Such constraints in the real world have implications for normative frameworks for application of risk analysis and management, such as guidelines, standards and regulations. When should risk analysis be carried out before a decision is made, what form should it take and how should it be documented? With regard to the decision, additional questions arise: How should alternative attributes be valued? How should uncertainty be valued? Obviously, different actors have different roles in risk management.

The roles and the character of risk handling are closely linked to the decision settings. We present a typology of decision settings, paying special attention to constraints and the potential for risk analysis and management. The classification is based on two dimensions: closeness to hazard and level of authority. It identifies decision settings that are typical for certain groups of actors and it discusses the appropriate constraints. It considers the implications of these constraints for decision-makers or actors with respect to risk analysis and management, and it shows the need for interaction among actors in different decision settings. There is a brief discussion on some normative ideas about groups of actors, their roles, responsibilities and interactions. Although the discussion is based on categories of quite stereotypical actors, we believe the classifications provide some insight into the limitations and the potential for risk management in different decision contexts.

Characterizations of decision settings

Figure 5.6 presents the two-dimensional taxonomy for categorizing decision settings. We think of proximity to hazard primarily in terms of physical distance and time. This implies that pilots, offshore platform superintendents or aircraft line maintenance personnel usually find themselves at the sharp end, i.e. close to the hazard source. Designers, planners, analysts and regulatory institutions typically operate at the blunt end. Some actors may be operationally close to the hazard source, even though they are physically remote, for instance air traffic

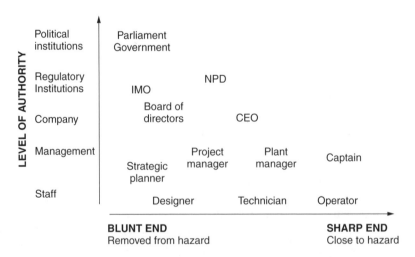

Figure 5.6 Two dimensions for characterizing settings for safety-related decision-making: IMO = International Maritime Organization, NPD = Norwegian Petroleum Directorate, CEO = chief executive officer

control operators or centralized train control operators. We will consider these actors as belonging to the sharp end, even though they are less vulnerable in the case of an accident. Actors at the sharp end are mostly event driven and thus operate within a shorter time horizon for most of the time. We also expect actors at the sharp end to have more updated and detailed hands-on knowledge of the system they operate than actors at the blunt end.

Level of authority is conceived primarily in formal terms. Actor A has a higher level of authority than actor B if actor A is entitled to give directives, orders or instructions to actor B but not vice versa. This does not necessarily imply that actor B is unable to exert power over actor A. Company executives may, for instance, work through political channels to exert pressure on a regulatory institution and influence standards and regulations.

The conditions under which actors make decisions strongly influence the decision processes which lead up to the decisions or to the way action is taken. We thus expect decision criteria, procedures and outcomes to be related to (1) how close an actor or decision forum is to the hazard and (2) the level of authority of the actor or forum. These relationships are complex, since decision-makers also adapt to circumstances not covered by these two dimensions. But even a grossly simplified model of these relationships may be helpful in sensitizing us to the way decision-makers adapt to their setting. Figure 5.7 shows a classification scheme based on five distinct decision settings.

The decision settings are characterized by typical contingencies and constraints, influencing the manner in which decisions are taken, including decision criteria, processes and limitations. We will consider the decision classes one by one. The constraints governing actors in a decision setting obviously impact their ability to analyse the outcome of alternative actions and/or to assess or deal individually with risk for each decision. When reviewing the various decision classes, we will also discuss the implications for risk analysis and management, and how risk can be dealt with in an appropriate manner, acknowledging that not

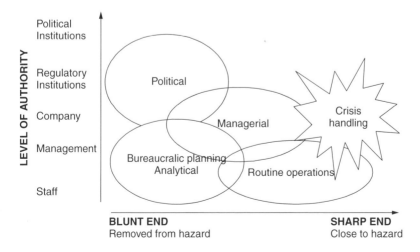

Figure 5.7 Classification of decision settings

all actors can collect information and model the world in detail before making a decision.

Routine operations

Let us first view decision-making in an operational environment, characterized by the sharp end and low to medium authority. In this setting (and possibly also in the others), action is not always the result of decisions, in the sense of conscious deliberations or analysis and choice of action. More detailed understanding of how information is processed by humans to produce courses of action in such settings can be found in the literature about human–machine interaction. The most common such setting is the three modes of activity generation from received information: skill based, rule based and knowledge based.

Skill-based behaviour is characterized by direct interaction between humans and their environment in an automated, feedforward control mode. It differs from rule-based or knowledge-based behaviour in that it does not relate to a 'problem', but translates information or cues through a mental model (e.g. experience of successful responses to inputs) into actions. Skill-based behaviour, in this sense, precedes a potential problem.

Rule-based behaviour relates to a problem in a standard 'if X takes a certain value x, then apply action d', rule-type manner. It relies on a repertoire of rules embedded in the decision-maker or the actor. In this sense it is a problem-solving activity; information is related to the presence of a problem. For successful application of a rule-based strategy, it is characteristic that the problem encountered is matched by an adequate rule. Otherwise the output of applying a rule will not be appropriate and it will lead to a hazard. Both skill and rules are generated through induction from specific experience and mental modelling to generalizations about appropriate reactions. Skills and rules can be conceptualized as pre-programmed solutions and contingency plans. Both cases generate a more or less automated response to changes in an observed world.

Knowledge-based behaviour in operational decision settings occurs when a problem is not addressed by the rule inventory, or when rules are broadly defined. It is a different form of problem solving than rule-based action as it involves analytic processes and prediction. Contrary to rule-based problem solving, knowledge-based problem solving is characteristic for situations where the problem is not well defined beforehand. In this categorization of behaviour, knowledge-based behaviour most closely resembles the classical picture of decision-making as problem solving.

The relative frequency of erroneous behaviour observed using rules or skills is low, whereas for knowledge-based decisions it is high.

Rules can be implicit and systems can have implicit reliance on rules. If safety relies on application of skills and rules, they often need to be formalized. In heavily regulated environments, e.g. aviation, reliance on explicit rules is strong. As a result, such operations tend to have competence requirements, rules and instructions that are more stringent and elaborate than those for less critical operations. For example, aircraft mechanics are subject to detailed personal competence and training requirements. Their work is performed in accordance with strict plans and detailed work instructions. Many potentially hazardous observations are listed in the documentation and accompanied by clear rules to follow.

Generally speaking, the operator will not refer to a model to make predictions about the effect on higher-level attributes; they will not be uncertain about these. The operator has observed a value x, which is certain. As long as there exists a rule – which is deterministic – uncertainty is not an issue for the decision-maker at the sharp end. Formalized, knowledge-based action in risk-sensitive environments will involve risk analysis. There exist examples of such formalisms, such as Safe Job Analysis, practised in the offshore industry, but even if such a formalism is adopted, important safety issues are often missed.

This does not imply that risk is not an issue at the sharp end. It is only a recognition of the fact that sharp-end behaviour is governed by responses to sensory inputs, which are predetermined and assume determinism in the relation between action and response. From the viewpoint of organizational risk management, it identifies the need for risk and uncertainty assessment elsewhere. Consequences of alternative decisions in response system behaviour need to be assessed beforehand, and strategies or detailed rules for behaviour need to be 'pre-programmed'. The 'elsewhere' can be viewed as a design assessment context. This is a typical blunt-end setting, where the available timescales and resources allow data collection and analysis.

The ideas presented here do not imply that such analyses have to be performed by a completely different category of people. The process of designing or pre-programming appropriate responses or decisions, depend on experience transfer from the sharp-end operational knowledge base. It appears quite sensible, even mandatory, that personnel from operations are involved in the risk analysis and pre-programming of decisions.

There are, however, practical limits to pre-programming of responses in complex and dynamic work environments. It may not be feasible to foresee all contingencies, and sharp-end personnel may not accept being pre-programmed by outsiders. In these situations a more sensible approach may be to provide operators with information on the boundaries of safe performance. The point is not to specify how the operator is to perform the job, but rather to show the boundary between safe and unsafe ways to do the job, see Rasmussen (1997).

Management

Management decisions, in the sense of unprogrammed decisions, can be associated with actors and decision settings at a high level of organizational authority and at the same time be somewhat removed from the sources of hazard. Examples are company boards, executives and senior managers or directors. Managers at this level could have typically up to 50 active problems to deal with at any given time. Studies of decision behaviour show that these actors, constrained by their information processing capacity, will often apply a satisficing strategy when making decisions. This implies that they will look for a decision option, which is good enough according to some aspiration level, see Section 5.1.3. Managers make many decisions without reference to anticipated consequences, but in accordance with rules and codes of conduct. This is seen as a simplification of decision-making based on successful previous applications. However, decisions involving major risks cannot be dealt with on the basis of prior experience. Rules of conduct for such decisions must therefore refer to uncertainty about future events, i.e. risk, which cannot be deduced solely from historical experience, as often that experience does not exist or is rather limited. For problems which involve large risks, managers will often choose to delegate all responsibility for the design phase to analytical functions; here 'design phase' means development of alternatives, analysis of consequence and risk, and development of a recommendation for a decision. Analytical functions can be interpreted as actors in a less exposed decision setting and at lower levels of authority. This coincides with an analytical, bureaucratic decision setting, see Figure 5.7. Decision-makers will retain the authority to approve a decision.

When risk analysis is carried out, the management decision-maker's risk assessment involves a more or less detailed assessment of the results of the risk and uncertainty analysis prepared by the experts and analysts. In our terminology this would coincide with a review of the predictions, the associated uncertainty assessments and relevant background information. Also if a formal decision analysis, for example a cost-benefit analysis, is performed, there is a need for a review and judgement process to choose the best decision alternative; see Section 5.1 for a more detailed discussion. Although many managers would apply a satisficing regime and use off-the-shelf standards in many situations, there is now wide acceptance for using a risk-based (informed) approach in situations involving high consequences and large uncertainties. Have a look at the classification system described in Section 5.3.1.

Political

Governmental and governmental agency decision-making is reflected in laws and regulations. Such decision actors or forums deal at high levels of authority and are far removed from safety hazard sources. The dominating decision-making processes in these settings are political or negotiative, supported by bureaucratic processing. The dominating constraint on these processes is conflicts of interest among stakeholders. The dominant decision criterion is thus to obtain the degree of consensus necessary to conclude the decision process. Such decisions should be seen less as solutions to well-defined problems and more as results of compromise, conflict and confusion through bargaining among actors with diverse interests. Many major decisions in national and international standardization forums (e.g. the International Organization for Standardization) and industrial organizations (e.g. the International Civil Aviation Organization and the International Maritime Organization) are made in this decision mode, in a discursive manner, similar to political decisions. With consensus as a major, albeit implicit, decision criterion, it is not meaningful to talk about optimal decisions in a conventional sense. The 'consensus' is part of the 'optimality' criterion. Moreover, changing coalitions may lead to inconsistencies of preferences with time.

We have assigned highly structured bureaucratic and political processes as well as open-ended or even chaotic political processes to a single class because bureaucratic and political decision processes are often tightly interwoven in practice. Political decisions are usually prepared and implemented by bureaucracies, and bureaucratic decisions may be appealed to political forums or deflected by actors working through political channels.

Due to the difficulties in achieving consensus on major changes from an existing platform, many political and bureaucratic decision processes come close to the so-called incremental muddling through paradigm (or successive limited comparisons) in which the actors build policy gradually through minor decisions based on limited analysis. In many cases such a process is not possible. Politicians need to make a number of far-reaching decisions, locally, regionally and globally. And looking at our parliaments, we see that politicians do in fact make a number of these decisions every year.

Uncertainty and risk analyses are requisite instruments in political decision-making. They are designed to support the political decisions by assessing consequences for alternative decision options and evaluation of consequences and risk against presumptive values and preferences.

Uncertainty and risk assessment should have an important place in informing public policy-makers (decision-makers). As for managerial decisions, the decision-makers should be informed about predicted consequences and the risk and uncertainty assessments. Considering the common lack of agreement by the political actors regarding the importance put on issues and objectives, care should be shown when using formal decision analysis. Such analyses should be used as decision aids, stressing that the value judgements adopted are used to produce insights and not hard recommendations.

Analytical or bureaucratic

In blunt-end settings, remote from immediate hazard and with no direct executive authority, we find functions like design, engineering and planning, as well as controlling and analytical functions. Actors in such functions are usually not forced to make decisions at the pace of executives. Their resources for information processing (e.g. time, calculation tools, data) tend to be relatively abundant. This often allows them to seek decision options, analyse and evaluate them and find the alternative that optimizes some criterion (e.g. NPV in a cost-benefit analysis) under the given constraints. The groups of actors and organizational functions falling into this setting are large and heterogeneous with respect to the nature of work and decisions. For some, the focus will be to make routine decisions, very similar to those described under operations, but more detached from hazards. Other functions are more supporting functions for decisions at higher levels.

We see three areas of involvement in risk management and decision-making for actors in this decision setting: (1) decisions made on the actors' own account, (2) provision of decision aid to decision-makers at higher authority levels or other actors inside the same category (e.g. analysts to designers) and (3) risk analysis and pre-programming of decision rules for sharp-end functions.

(1) Although actors in the bureaucratic domain tend to have more time, information and information processing resources than actors in other domains, this does not imply that optimization will be the dominant decision mode. Decisions belonging to category (1) will on many occasions be made by following rules of code, or through satisficing against predetermined criteria. A designer has to relate to constraints of cost, weight, functionality, production limitations, reliable operation, etc. The designer and his or her manager, normally a middle manager with limited overall authority, can be expected to analyse and judge one alternative against the local requirements. Seldom will an overall optimization take place. For certain types of equipment, such as critical aircraft components, a risk (reliability) analysis will be performed for the component and its function, which feeds into a global safety assessment for an aircraft as a whole. In these cases we find close resemblance to an idealized risk analysis. In terms of a decision analysis, the setting is more of a satisficing regime than an optimization regime. Optimization requires parallel analysis and evaluation of relevant alternatives, i.e. more than one alternative. Often all but one alternative would have been eliminated before performing a detailed assessment.

(2) Risk and decision analysis as an aid to executive decision-making can take various forms of detail and completeness. The analyst receives an assignment from a manager with higher authority. The task is to recommend the best possible solution to a problem. This is a setting typical of more strategic decision analysis. The executive has defined the problem. The process of identifying alternatives, analysing them with respect to their consequences and risk, evaluating them and recommending a choice on this basis resembles the classical structure of decision analysis. The tasks of the analyst are (a) with more or less involvement from the decision maker, define relevant affected objectives; (b) establish a set of alternative decisions or options to be assessed; (c) with assistance from databases and experts, for each decision alternative collect data and information to be used; (d) establish some form of model (fault tree, cause-consequence tree, etc.) relating knowledge at a lower level to expressions of consequences and risk at a higher level. Now a recommendation for a decision could be made on a direct evaluation and heuristic choice, based on predictions and risk statements, or more formal methods could be employed. Some standards for decision-making involving risk encourage cost-benefit analysis in the ALARP region. In this case, explicit value trade-offs and/or expected monetary values of consequences would be required.

(3) Personnel dealing with sharp-end situations tend to apply pre-programmed skills and rules in dealing with system feedback and problems. This implies that a set of contingent decision rules to deal with possible system states needs to be developed. This can be achieved after a prior risk analysis. On the basis of undesirable outcomes, one needs to assess which observations could produce these outcomes. This can be done by using fault tree techniques, for example. Once a set of limiting values for the observations has been defined, rules can be assessed to reduce the risk of a negative outcome. In addition to the specification of rules, the product of these exercises should be documentation of

the assumptions used in the analysis and the criteria used in determining the rule set. An important element of the blunt-end pre-programming, then, is the continued experience feedback and updating of knowledge, risk and uncertainty and, accordingly, the rules. Experience can then be compared to the predicted consequences and the risk statements.

Crises and emergency management

Crisis and emergency are given many meanings in the literature, ranging from a situation which is not manageable inside normal planning and processing routines, via presence of serious threats that require prompt action, to extremely dynamic situations with major consequences, such as fires. These situations have in common that they relate to an environment evolving dynamically with serious, but uncertain consequences. Here we focus on situations with a high degree of seriousness. Decision-making is mainly concerned with limiting negative consequences. During crises different patterns of decision-making are observed and are required. The rate of information is often high, the time constraints are narrow, the options may not be obvious and the consequences of an action will be uncertain. Decision-makers who normally perform in a blunt-end manner, perform under extreme hazard exposure. A decision-maker faced with a crisis needs not only to find a way to avoid adverse outcomes. He also needs to limit anxiety and stress to a level that is tolerable and compatible with efficient coping. Unaided, the likelihood of inadequate decisions is high.

Appropriate behaviour in emergency and crisis settings obviously depends on contingency planning and emergency training. Because we are dealing with situations for which there usually exists little or no direct experience and which develop highly dynamically, this type of planning requires prior risk analysis. The purpose of the risk analysis in these cases is not to support a specific decision, because the problem is not current or known in detail. The purpose is rather to identify generic decisions and tie them to certain classes of situation. An example could be a procedure to perform an emergency landing of a helicopter in the event of sudden, heavy vibrations. No specific causal analysis is used to support such a decision; no specific analysis of the direct effect of the vibration supports this decision. The procedure is deduced from the knowledge that a number of critical failures could produce heavy vibration (the class of failures producing vibrations), and an effective decision to mitigate this risk is to perform an emergency landing. Crisis management cannot be strongly linked to a specific level of authority. In a crisis the roles and authority of an individual can change. Depending on the severity of the crisis, functions at practically all levels of authority can become involved in decision-making.

Emergencies are associated with high consequence contingencies and low probabilities. One could consider them a form of residue of the risk assessment. Because they are not dealt with in the normal risk decision-making and management processes, they require a different approach. The purpose of risk analysis and decision analysis in the case of crisis and emergency management is (a) to identify critical situations to a degree possible, (b) to devise generic strategies as a planning basis, (c) to predetermine roles and responsibilities in the case of emergency and (d) to allocate resources for emergencies. Planners and analysts should convince the manager and provide them with a plan for an immediately available course of action worked out under calmer circumstances. This is similar to the pre-programming of rules in support of operator environments. Professional analysts should have a role in crisis situations as providers of real-time analysis, to offload the managers' need for information processing. Such work sharing is advocated even if the analysis would have to be quick and dirty.

Interaction between classes of decision setting: roles and responsibilities

It is clear that risk management requires close interactions among classes of decision settings. Sometimes these can overlap with specific organizational functions but they are not always identified as such. For example, a senior manager can be seen in a strategic management function, but under certain circumstances he can also perform as a crisis manager. The two cases would represent radically different constraints and, accordingly, the mode of decision-making would be expected to vary. Constraints of many settings in which decisions affecting risk are taken do not allow for formal analysis. Distribution of roles and responsibilities between analytical functions, operational functions and executive management functions can be understood by keeping in mind the decision settings and modes.

Figure 5.8 sketches a framework for the different roles, responsibilities and relations. Higher-authority and sharp-end actors provide a knowledge base and a frame of legal, moral and commercial values.

Political institutions and standardization agencies process public norms and values through different forms of discourse and decisions on laws, regulation

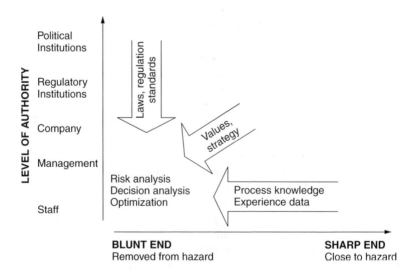

Figure 5.8 Influences from high-authority and sharp-end decision settings on analytical processes

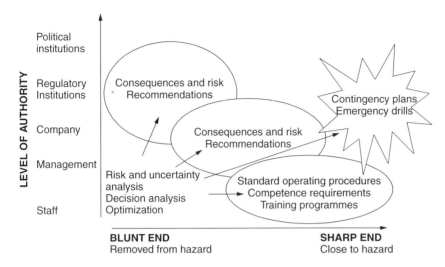

Figure 5.9 Analytical support of high-authority and sharp-end decision settings

or standards. These form part of the background and influence organizational assessment of risk. Executive management positions of companies express values and strategic priorities through strategy documents, budget guidelines and a variety of formal and informal instructions and messages. These form references for analysis and evaluations by analytical functions. Operational environments provide updated process knowledge and experience data, which serve as input to analytical processes through reporting systems, database records and informal communication. The analytical function processes these inputs and information through model building, drawing inferences about prediction and risk and, possibly, some form of optimization. The product or output of the analytical function is largely support and pre-programming of decisions for decision settings that do not favour formal analysis. These principles are stylized in Figure 5.9.

For the political setting and the managerial setting, the output would consist of predicted consequences and risk, and in some cases it would include recommendations for decision. For operational environments the analysis would provide skill or competence requirements and standard operating procedures (e.g. operation manuals, maintenance manuals, troubleshooting manuals). Contingency planning requirements should be identified for all settings, including emergency procedures, contingency measures and resources, and requirements for emergency practices. In order to have an impact, risk analysts need to understand the constraints facing decision-makers in other settings, and the strategies used by decision-makers to cope with these constraints.

From the discussions here it seems apparent that risk and uncertainty are dealt with, managed, through interaction and communication among a large number of actors. The rather rudimentary and static picture drawn here is limited by its generality. More detailed networks of interactions could be shown for more specific societal areas, industries, life-cycle phases and organizations. But we will not go further into this here. What appears clear is that for the formal interactions and processing of risk, we require a common understanding and a common terminology.

BIBLIOGRAPHIC NOTES

Section 5.1 is based on Aven and Kørte (2003). For overviews and discussions of formal decision analysis, see Bedford and Cooke (2001), Clemen (1996), Watson and Buede (1987) and Bell et al. (1988). An excellent introduction to the Bayesian utility paradigm is given by Lindley (1985). The structure of the decision-making process in Section 5.1 largely overlaps with the ideas of Hertz and Thomas (1983).

We have used decision analysis as an aid for providing insight into the decision-making process rather than recommending hard decisions. This way of using decision analysis is in line with for example French and Insua (2000) and Watson and Buede (1987), but does not seem to be held by most exponents of the theory. The pioneers of the economic decision-making school, and later the Bayesian decision-making theorists, seem to have a thinking where the decision should be specified by the result of the decision analysis.

Implicit values of a statistical life are reported in Tenga et al. (1995) and Ramsberg and Sjöberg (1997); see also Bedford and Cooke (2001: 363).

Multi-attribute utility theory is reviewed by Clemen (1996) and Keeney and Raiffa (1976), among others.

We have been strongly inspired by the ideas of Watson and Buede (1987), in particular on rationality and group decision-making. This book gives an overview of thinking on decision-making in organizations. A key author here is H.A. Simon, who introduced the concepts of bounded rationality and satisficing decision procedures, see Simon (1957a, 1957b). See also March and Simon (1958), Cyert and March (1992) and Bell et al. (1988). Several textbooks cover decision-making in organizations, e.g. Bell et al. (1988), French and Insua (2000), Mintzberg (1973) and Allison and Zelikow (1999).

The expected utility approach is established for an individual decision maker. No coherent approach exists for making decision by a group. K.J. Arrow proved in 1951 that it is impossible to establish a method for group decision-making that is both rational and democratic, based on four reasonable conditions that he felt should be fulfilled by a procedure for determining a group's preferences between a set of alternatives, as a function of the preferences of the group members (Arrow 1951). A considerable literature has been spawned from Arrow's result, endeavouring to rescue the hope of creating satisfactory procedures for aggregating views in a group. But Arrow's result remains as strong as ever. See French and Insua (2000: 108) and Watson and Buede (1987: 108).

The general discussion on acceptable risk problems in Section 5.2.3 is based on Ale (1999). The ALARP principle is reviewed by Aven and Pitblado (1998), Pape (1997), UKOOA (1999) and NORSOK (2001), among others. A good overview on the theory of risk perception is given by Okrent and Pidgeon (1998). The example of Section 5.2.4 is taken from Aven (1992). The health risk example of Section 5.2.5 is partly based on Natvig (1997). The warranty example in Section 5.2.7 is inspired by Singpurwalla (2000). The risk assessment approach in Section 5.2.8 is partly based on Vinnem *et al.* (2002).

The concept of deliberation in Section 5.1.3 is from Stern and Fineberg (1996).

The classification scheme in Section 5.3.1 is taken from Kristensen *et al.* (2003). The structure of this scheme is inspired by and partly based on Klinke and Renn (2001). Their framework is based on a classical view of risk as presented in Chapter 2. The accident risk classification scheme in Section 5.3.2 is based on Kørte *et al.* (2002). This paper extends a taxonomy of decision-makers introduced by Rosness (Rosness and Hovden 2001). Several researchers have earlier contrasted the position of sharp-end personnel that directly operate hazardous systems to the position of managers and designers at the blunt end, who strongly influence the tasks and working conditions of sharp-end personnel; see Reason (1997). To understand how information is processed by humans to produce courses of actions in an operational environment, consult the literature about human–machine interaction, e.g. Rasmussen (1986), and studies of human reliability and error, e.g. Reason (1990).

Other similar classification schemes have also been presented in the literature; see Rasmussen (1997).

The muddling through paradigm is described by Lindblom (1995); see also Schulman (1995).

For further discussion on crisis and emergency management, see Mintzberg (1973), Janis and Mann (1977), Samurcay and Rogalski (1991), Klein and Crandall (1995) and Rasmussen (1991).

6

Summary and Conclusions

This chapter summarizes the main conclusions made in this book, with reference to the relevant pages for the detailed presentation and discussion. When planning, conducting and using risk analysis, we believe that the following points should be adopted as a general guide:

1. Focus on quantities expressing states of the 'world', i.e. quantities of the physical reality or nature that are unknown at the time of the analysis but will, if the system being analysed is actually implemented, take some value in the future, and possibly become known. We refer to these quantities as *observable* quantities. (p. 48)
2. The observable quantities are predicted. Uncertainty related to the observable quantities is expressed by means of probabilities. This uncertainty is *epistemic*, i.e. a result of lack of knowledge. We cannot recommend the common procedure of always thinking of underlying physical phenomena producing some 'true' distributions. The starting point is that we lack knowledge about the observable quantities and we use probabilities to express this lack of knowledge. (p. 48)
3. Probabilities are based on a comparison with an urn model (or a probability wheel) – when the analyst assigns a probability of 10%, say, it means that his uncertainty is the same as drawing a favourable ball from an urn with 10% favourable balls under standard experimental conditions. In principle it is meaningless to speak about the correctness of an assigned probability, as a probability in our setting is a subjective measure. However, in some cases, comparisons can be made with observations of the observable quantities, but at the point of analysis the probabilities cannot be fully 'verified' as a probability expresses uncertainty about an observable quantity viewed by the analyst. What can be done is to review the background information used as the rationale for the assignment, but in most cases it would not be possible to explicitly document all the transformation steps from this background information to the assigned probability. (p. 64)

4. Training of risk analysts and experts should make them aware of factors, including heuristics, that influence probability assignments.
5. To avoid unwanted variability, standardization of some probability assignments are required when using risk analysis in a company, for example. In general, consensus on probabilities is desirable. (p. 68)
6. Probabilities are always conditioned on background information, and this information should be reported with the specified probabilities. (p. 50)
7. Models in a risk analysis context are deterministic functions linking observable quantities on different levels of detail. The models are simplified representations of the world. It is essential to discuss the goodness or appropriateness of the models to be used in a risk analysis, but the term 'model uncertainty' has no meaning in our framework. The models used are part of the background information. (pp. 48, 68)
8. A chance defined by the proportion of an infinite or very large population of units having a certain property, is an observable quantity. (p. 79)
9. Different techniques exist to assess uncertainty and specify a probability for an observable quantity:

   - Modelling expresses the observable quantity as a function of a number of other observable quantities. It is often easier to assess uncertainties of observable quantities on this more detailed level. Modelling is used to get insight into the system performance, to identify the risk contributors and see the effect of changes. (p. 68)
   - If historical data are available, classical statistical methods can be used as a basis for assigning the probabilities. To use this approach, the observational data must be judged relevant and the number of observation must be quite large. (p. 72)
   - Analyst judgment using all sources of information is commonly adopted when data are absent or are only partially relevant to the assessment endpoint. (p. 73)
   - Formal expert elicitation should be used when few data are available and the assignments are likely to be scrutinized. (p. 74)
   - Use a probability distribution class, e.g. the Poisson distribution, with fixed parameter values, when the background information is fairly strong. (p. 81)
   - A full Bayesian analysis with specification of a prior distribution should be used when seeking a mechanical and coherent updating procedure for incorporating new information. Informative prior distributions should preferably be used. (p. 72)
   - A full Bayesian analysis with specification of a prior distribution could also be used when little information is available. Meaningful interpretations of the parameters and the prior (posterior) distribution should be given. (p. 82)

10. The risk analyses present predictions and uncertainty assessments of observable quantities. They provide decision support. (p. 98)

SUMMARY AND CONCLUSIONS

11. Formal decision analyses, such as multi-attribute analyses, cost-benefit analyses and utility-based analyses, also provide decision support but not hard decisions. The analyses need to be put into a wider decision-making context, which we call a managerial review and judgment process, and this process results in a decision. (p. 97)
12. Explicit trade-offs between the various attributes need not always be performed to provide a good basis for decision. (p. 105)
13. In a cost-benefit analysis there exist no objective reference values for the statistical cost of a life. (p. 101)
14. In most cases, multi-attribute analyses and versions of cost-benefit analyses are rather easy to conduct compared with utility-based analyses. (p. 104)

Appendix A

Basic Theory of Probability and Statistics

This appendix gives a brief summary of basic probability theory and statistical inference. See the bibliographic notes for an overview of some key textbooks and papers in the field.

A.1 PROBABILITY THEORY

Probabilities are used when considering future events with more than one possible outcome. In a given situation only one of these outcomes will occur; in advance we cannot say which. Such situations are often called stochastic, or random, as opposed to deterministic situations where the outcome is determined in advance.

In the following we will give a precise definition of what we mean by a probability and the rules that apply for dealing with probabilities.

A.1.1 Types of Probabilities

The probability of an event A, $P(A)$, can be defined in different ways. It is common to distinguish between three types of probabilities, or more precisely, three conceptual interpretations:

- classical;
- relative frequency;
- subjective.

The classical interpretation applies only in situations with a finite number of outcomes that are equally likely to occur. According to the classical interpretation, we have

$$P(A) = \frac{\text{Number of outcomes resulting in } A}{\text{Total number of outcomes}}.$$

As an example consider the tossing of a die. Here P(the dice shows 2) $= 1/6$ since there are six possible outcomes which are equally likely to appear.

Following the relative frequency interpretation, probability is defined as the fraction of times the event A occurs if the situation considered were repeated (in real life or hypothetically) an infinite number of times. If an experiment is performed n times and the event A occurs n_A times, then

$$P(A) = \lim_{n \to \infty} \frac{n_A}{n},$$

i.e. the probability of the event A is the limit of the fraction of the number of times event A occurs when the number of experiments increases to infinity. Note that a classical interpreted probability is equal to a relative frequency interpreted probability. In our die example the proportion of dies showing 2 is 1/6 in the long run, hence the relative frequency interpreted probability is 1/6.

It is common also to refer to the relative frequency interpretation as the classical interpretation, and we adopt that convention in this book.

In most real-life situations, the relative frequency interpreted probability is unknown and has to be estimated from experience data. Here is an example.

Example A.1 *We consider a fire detector of a certain type K. The function of the detector is to raise the alarm at a fire. Let A denote the event 'the detector does not raise the alarm at a fire'. To find $P(A)$, assume that tests of n detectors of type K have been carried out and the number of detectors that are not functioning, n_A, is registered. As n increases, the fraction n_A/n will be approximately constant and approach a certain value (this fact is called the strong law of large numbers). This limiting value is called the probability of A, $P(A)$. If $n = 10\,000$ and we have observed $n_A = 50$, then $P(A) \approx 50/10\,000 = 5/1000 = 0.005$ (0.5%). Note that a probability is by definition a number between 0 and 1, but the quantity is also often expressed as a percentage.*

The relative frequency interpretation is discussed in more depth in Chapter 2.

In the subjective interpretation, $P(A)$ is a subjective measure of uncertainty. This means that we (who assign the probability) compare the uncertainty of event A occurring with drawing a favourable ball from an urn having $P(A) \times 100\%$ favourable balls under standard experimental conditions. This means that we have the same degree of belief in the event A occurring as drawing a favourable ball from an urn with $P(A) \times 100\%$ favourable balls. Subjective probabilities are thoroughly discussed in Chapter 4.

All probabilities are conditioned on some background information K, say. Thus a more precise way of writing the probability $P(A)$ is $P(A|K)$, which is the common way of expressing a conditional probability. To simplify the notation, we normally omit the K. This should not cause any problem as long as the background information is fixed throughout the argument.

BASIC THEORY OF PROBABILITY AND STATISTICS

A.1.2 Probability Rules

Before we summarize some basic rules for probabilities, here is an overview of some definitions from set theory. (The probability interpretations are given in parentheses).

Definitions

The empty set	$\emptyset$	A set with no elements (outcomes) (impossible event)
Basic set (sample space)	S	A set comprising all the elements we are considering (a certain event)
Subsets	$A \subset B$	A is a subset of B, i.e. each element of A is also an element of B (if the event A occurs, then the event B will also occur)
Equality	$A = B$	A has the same elements as B (if the event A occurs, then also the event B occurs, and vice versa)
Union	$A \cup B$	$A \cup B$ includes all the elements of A and B ($A \cup B$ occurs if either A or B occurs (or both), i.e. at least one of the events occur)
Intersection	$A \cap B$	$A \cap B$ includes only elements which are common for A and B ($A \cap B$ occurs if both A and B occur)
Disjoint sets	$A \cap B = \emptyset$	The sets have no common elements (A and B cannot both occur)
Difference	$A - B$	$A - B$ includes all elements of A that are not elements in B ($A - B$ occurs if A occurs but B does not occur)
Complement	$\overline{A}$	$\overline{A}$ includes all elements of S that are not elements of A ($\overline{A}$ occurs if A does not occur)

Some of these definitions are illustrated by Venn diagrams in Figure A.1. The following fundamental rules apply:

$$A \cup B = B \cup A,$$
$$A \cap B = B \cap A,$$
$$(A \cup B) \cup C = A \cup (B \cup C) = A \cup B \cup C,$$
$$(A \cap B) \cap C = A \cap (B \cap C) = A \cap B \cap C,$$

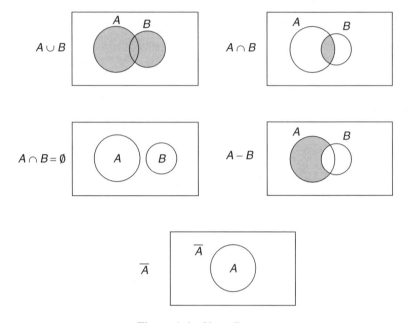

Figure A.1 Venn diagrams

$$A \cap (B \cup C) = (A \cap B) \cup (A \cap C),$$
$$A \cup (B \cap C) = (A \cup B) \cap (A \cup C),$$
$$\overline{A \cup B} = \overline{A} \cap \overline{B},$$
$$\overline{A \cap B} = \overline{A} \cup \overline{B},$$
$$A \cup \overline{A} = S.$$

Modern probability theory is not based on any particular interpretation of probability, although its standard language is best suited to the classical and relative frequency interpretations. Throughout the presentation we highlight differences between the relative frequency interpretation and the subjective interpretation. The starting point is a set of rules, known as Kolmogorov's axioms, that have to be satisfied. Let $A, A_1, A_2, \ldots$ denote events in the sample space S. For Example A.1 the sample space comprises the events 'the detector raises the alarm at a fire' and 'the detector does not raise the alarm at a fire'.

The following probability axioms are assumed to hold:

- $0 \leq P(A)$,
- $P(S) = 1$,
- $P(A_1 \cup A_2 \cup \cdots) = P(A_1) + P(A_2) + \cdots +$,
  if $A_i \cap A_j = \emptyset$ for all i and j, $i \neq j$.

BASIC THEORY OF PROBABILITY AND STATISTICS 153

Based on these axioms it is possible to deduce the following probability rules:

$$P(\overline{A}) = 1 - P(A),$$

$$P(A_1 \cup A_2) = P(A_1) + P(A_2) - P(A_1 \cap A_2),$$

$$A_1 \subset A_2 \Rightarrow P(A_1) \leq P(A_2).$$

Conditional probabilities

The conditional probability of the event B given the event A is denoted $P(B|A)$. As an example, consider two components and let A denote the event 'component 1 is not functioning' and let B denote the event 'component 2 is not functioning'. The conditional probability $P(B|A)$ expresses the probability that component 2 is not functioning when it is known that component 1 is not functioning.

The conditional probability $P(B|A)$ is defined by

$$P(B|A) = \frac{P(B \cap A)}{P(A)}, \tag{A.1}$$

whenever $P(A) > 0$. Calculation rules for standard unconditional probabilities also apply to conditional probabilities. From (A.1) we see that

$$P(A \cap B) = P(B \mid A) P(A).$$

More generally we have

$$P(A_1 \cap A_2 \cap \cdots \cap A_n) = P(A_1) P(A_2|A_1) \cdots P(A_n|A_1 \cap A_2 \cap \cdots \cap A_{n-1}).$$

Some other important rules including conditional probabilities are:

$$P(B \mid A) = \frac{P(A \mid B) P(B)}{P(A)}; \tag{A.2}$$

if $\cup_{i=1}^{r} A_i = S$ and $A_i \cap A_j = \emptyset$, $i \neq j$, then

$$P(B) = \sum_{i=1}^{r} P(B \cap A_i) = \sum_{i=1}^{r} P(B \mid A_i) P(A_i). \tag{A.3}$$

Equation (A.2) is known as Bayes' theorem and equation (A.3) as the law of total probability.

Independence

Two events, A and B, are said to be independent if the occurrence or non-occurrence of one does not change the occurrence probability of the other.

Mathematically, this means that

$$P(B \mid A) = P(B),$$

or equivalently

$$P(A \mid B) = P(A),$$

$$P(A \cap B) = P(A)\,P(B).$$

If A and B are independent, then A and $\overline{B}$ are also independent, as well as $\overline{A}$ and B, and $\overline{A}$ and $\overline{B}$. If the subjective probability interpretation is adopted, we say that A and B are *judged* independent.

The events $A_1, A_2, \ldots, A_n$ are (judged) independent if

$$P(A_{i_1} \cap A_{i_2} \cap \cdots \cap A_{i_r}) = \prod_{j=1}^{r} P(A_{i_j})$$

for any set of different indices $\{i_1, i_2, \ldots, i_r\}, r = 1, 2, \ldots, n$, taken from the set $\{1, 2, \ldots, n\}$.

Example A.2 *Refer back to Example A.1. Assume we have established that $P(A) = 0.005$, where A denotes the event 'the detector does not raise the alarm at a fire'. To reduce the probability of no alarm at a fire, we install two detectors. The problem is now to compute the probability of the following events:*

$B = $ *'No detectors are functioning at a fire',*

$C = $ *'At least one of the detectors is functioning at a fire'.*

To compute these probabilities, let A_i, $i = 1, 2$, denote the event 'detector i does not function at a fire'. Then $B = A_1 \cap A_2$ and $C = \overline{A}_1 \cup \overline{A}_2$. We know that $P(A_1) = P(A_2) = 0.005$, but this information is not sufficient for calculating $P(B)$ and $P(C)$. Assuming A_1 and A_2 are independent, we find that

$$P(B) = P(A_1 \cap A_2) = P(A_1)P(A_2) = 0.005^2 = 0.25 \times 10^{-4},$$

$$P(C) = P(\overline{A}_1 \cup \overline{A}_2) = 1 - P(A_1)P(A_2) = 0.999975.$$

Alternatively, we could have found $P(C)$ by

$$P(C) = P(\overline{A}_1) + P(\overline{A}_2) - P(\overline{A}_1)P(\overline{A}_2)$$

$$= 0.995 + 0.995 - 0.995^2 = 0.999975.$$

Given that at least one of the detectors does not function, what is the probability that detector 1 is not functioning? Intuitively, it is clear that this conditional probability will be approximately 50%. To show this formally, note that this probability can be expressed as $P(A_1 | A_1 \cup A_2)$. Use of various probability rules gives

$$P(A_1 \mid A_1 \cup A_2) = \frac{P[A_1 \cap (A_1 \cup A_2)]}{P(A_1 \cup A_2)} = \frac{P(A_1)}{P(A_1) + P(A_2) - P(A_1)P(A_2)}$$

$$= \frac{0.005}{0.005 + 0.005 - 0.005^2} \approx \frac{1}{2},$$

as expected.

A.1.3 Random Quantities (Random Variables)

In applications we often focus on one or more summarizing performance measures, in contrast to all possible outcomes. Let us return to the detector example. Assume that we are considering k detectors. We are primarily interested in the *number* of detectors that are not functioning, i.e. not raising the alarm. Let X denote this number. The value of X is uniquely given when the outcome of the 'experiment' is registered. If, for example, $k = 2$ and it is observed that detector 1 is functioning but not detector 2, then $X = 1$. Thus we may view X as a function from the sample space to the real numbers. We call such variables random variables or stochastic variables. If the subjective probability interpretation is adopted, it is common to refer to X as a random quantity. The word 'variable' is usually avoided as it gives the wrong impression that X varies. We will use the term 'random quantity' as the generic term and refer to random variables only when interpreting probability in a classical or relative frequency way.

Let in general X denote a random quantity and assume that X is discrete, i.e. it can only take a finite number of values or a countably infinite number of values. Let $P(X = x)$ denote the probability of the event '$X = x$', where x is one of the values X can take. We call the function $f(x) = P(X = x)$ the probability distribution of X, or simply the distribution of X.

In many applications we prefer to work with random quantities having continuous distributions, i.e. distributions characterized by a probability density $f(x)$ such that

$$P(a < X \leq b) = \int_a^b f(x)\,dx.$$

Thus if $b - a$ is small,

$$P(a < X \leq b) \approx f(x)(b - a).$$

Mean and variance of X

The *mean* or the *expected* value of X, EX, is defined as

$$EX = \sum_x x\, P(X = x).$$

From the definition we see that EX can be interpreted as the centre of mass of the distribution. Consider again the fire detector example. It follows from the strong law of large numbers that EX is approximately equal to the average number of detectors that are not functioning among the k, if we look at a large number of identical collections of k detectors. Hence the mean can also be interpreted as an average value.

The variance of X, $\text{Var}\,X$, is a measure of the spread or variability of the values of X around EX, and is defined by

$$\text{Var}\,X = \sum_x (x - EX)^2\, P(X = x).$$

The standard deviation of X is given by $\sqrt{\text{Var}\,X}$.

The mean and variance in the continuous case are defined by

$$EX = \int_{-\infty}^{\infty} x f(x)\,dx,$$

$$\text{Var}\,X = \int_{-\infty}^{\infty} (x - EX)^2 f(x)\,dx.$$

Independence

Let $X_1, X_2, \ldots, X_n$ denote n arbitrary random quantities. We say that these quantities are independent if

$$P(X_1 \le x_1 \cap X_2 \le x_2 \cap \cdots \cap X_n \le x_n) = \prod_{i=1}^{n} P(X_i \le x_i)$$

for all choice of $x_1, x_2, \ldots, x_n$. In a subjective probability context, independence means judged independence.

Exchangeability

Next we introduce the notion of exchangeability. Consider two discrete random quantities, X_1 and X_2. Then X_1 and X_2 are said to be exchangeable if for all values x_1 and x_2 that X_1 and X_2 can take, we have

$$P(X_1 = x_1 \text{ and } X_2 = x_2) = P(X_1 = x_2 \text{ and } X_2 = x_1);$$

that is, the assessed probabilities are unchanged (invariant) by switching (permuting) the indices.

More generally, random quantities $X_1, X_2, \ldots, X_n$ are exchangeable if their joint distribution is invariant under permutations of coordinates, i.e.

$$F(x_1, x_2, \ldots, x_n) = F(x_{r_1}, x_{r_2}, \ldots, x_{r_n}),$$

where F is a generic joint cumulative distribution for $X_1, X_2, \ldots, X_n$ and equality holds for all permutation vectors $(r_1, r_2, \ldots, r_n)$.

Exchangeability is weaker than independence because, in general, exchangeable random quantities are dependent. Independent random quantities having identical probability distributions are exchangeable, but not vice versa. In a subjective probability context, exchangeability means judged exchangeability.

BASIC THEORY OF PROBABILITY AND STATISTICS 157

Some rules for random quantities

Here are some important rules for the mean and variance (a and b are constants)

$$E(aX + b) = aEX + b,$$

$$EX \leq EY \text{ if } X \leq Y,$$

$$E(X_1 + X_2 + \cdots + X_n) = EX_1 + EX_2 + \cdots + EX_n,$$

$$\text{Var}(aX + b) = a^2 \text{Var } X,$$

$$Eh(X) = \begin{cases} \sum_x h(x) P(X = x) & \text{if } X \text{ is discrete,} \\ \int_{-\infty}^{\infty} h(x) f(x) \, dx & \text{if } X \text{ is continuous.} \end{cases}$$

If the X_i are independent, then

$$\text{Var}(X_1 + X_2 + \cdots + X_n) = \text{Var } X_1 + \text{Var } X_2 + \cdots + \text{Var } X_n.$$

In the general case

$$\text{Var}(X_1 + X_2 + \cdots + X_n) = \sum_{i=1}^{n} \text{Var } X_i + 2 \sum_{j<l} \text{Cov}(X_j, X_l),$$

where $\text{Cov}(X_j, X_l) = E(X_j - EX_j)(X_l - EX_l)$ is the covariance of X_j and X_l. The covariance is closely related to the correlation coefficient, ρ, defined by

$$\rho(X_j, X_l) = \text{Cov}(X_j, X_l)/(\sigma_{X_j} \sigma_{X_l}),$$

where σ_X is the standard deviation of X. The correlation coefficient ρ satisfies $\rho \in [-1, 1]$.

Conditional probability and expectation

One of the most useful concepts in probability theory is that of conditional probability and expectation. Let X and Y be two discrete random quantities. Then the conditional probability distribution of Y given that $X = x$ is

$$f(y|x) = P(Y = y|X = x) = \frac{P(Y = y, X = x)}{P(X = x)}$$

for all values such that $P(X = x) > 0$. The conditional expectation of Y given $X = x$ is defined by

$$E(Y|X = x) = \sum_y y f(y|x).$$

Similarly, we can define a conditional probability distribution of X and a conditional expectation for continuously distributed random quantities:

$$f(y|x) = f(y, x)/g(x),$$

$$E(Y|X = x) = \int_{-\infty}^{\infty} y f(y|x) \, dy,$$

where $f(y, x)$ is the joint density function for the random quantities Y and X, given by

$$P(a < Y \leq b, c < X \leq d) = \int_a^b \int_c^d f(y, x) \, dx \, dy,$$

and $g(x)$ is the probability density of X. Let $E(Y|X)$ denote the function of the random quantity X whose value at $X = x$ is $E(Y|X = x)$. Note that $E(Y|X)$ is itself a random quantity. Then it can be shown that

$$EY = EE(Y|X). \tag{A.4}$$

If Y is a discrete random quantity, then this equation states that

$$EY = \sum_x E(Y|X = x) P(X = x),$$

while if X is a continuous random quantity with density $g(x)$, then it states that

$$EY = \int_{-\infty}^{\infty} E(Y|X = x) g(x) \, dx.$$

If X_1 and X_2 are independent random quantities, having continuous distributions F_1 and F_2, respectively, the distribution of the sum, $Y = X_1 + X_2$, is given as

$$P(Y \leq y) = \int_{-\infty}^{\infty} F_1(y - x) f_2(x) \, dx,$$

where f_2 is the density of X_2. The analogous formula in the discrete case is

$$P(Y \leq y) = \sum_x F_1(y - x) P(X_2 = x).$$

These formulas follow by applying the rule (A.4) with Y replaced by the indicator, function, which is one if $Y \leq y$ and zero otherwise. The distribution of Y is known as the convolution of the distributions F_1 and F_2.

The strong law of large numbers

The following theorem, known as the strong law of large numbers, is one of the most well-known results in probability theory. It states that the average of a sequence of independent random quantities having the same distribution will, with probability one, converge to the mean of that distribution.

Theorem A.1 *Let $X_1, X_2, \ldots$ be a sequence of independent random quantities having a common distribution, and let $EX_i = \mu$. Then with probability one*

$$\frac{X_1 + X_2 + \cdots + X_n}{n} \to \mu \quad \text{as } n \to \infty.$$

BASIC THEORY OF PROBABILITY AND STATISTICS

A.1.4 Some Common Discrete Probability Distributions (Models)

Here are some common discrete distributions, often known as probability models, following a standard presentation in the classical setting. Section A.1.6 contains some comments on how to interpret and use these distributions in a framework based on subjective probabilities.

Binomial distribution

The binomial distribution is used in situations where a series of independent trials are performed, where each trial results in either success or failure. These trials are called Bernoulli trials. If p is the constant probability of success in a trial and if k is the number of trials, then the total number of successes, which we denote by X, is binomially distributed with parameters k and p, i.e.

$$P(X = x) = \binom{k}{x} p^x (1-p)^{k-x},$$

where the binomial coefficient $\binom{k}{x}$ is defined by

$$\binom{k}{x} = k!/x!(k-x)!.$$

Here $k! = 1 \times 2 \times 3 \times \cdots \times k$, etc. For a binomial distribution it can be shown that

$$EX = kp \quad \text{and} \quad \text{Var}\, X = kp(1-p).$$

In the fire detector example, X is binomially distributed with parameters k and $p = 0.995$. Note that if X is binomially distributed with parameters k and p, then $k - X$, which represents the number of failures, is binomially distributed with parameters k and $1 - p$.

Geometric distribution

The geometric distribution is closely related to the binomial distribution. Consider a series of independent Bernoulli trials with p denoting the probability of success. Then X, defined by the number of trials required until the first success, is geometrically distributed with parameter p, i.e.

$$P(X = x) = p(1-p)^{x-1}, x = 1, 2, \ldots.$$

For this distribution we have

$$EX = \frac{1}{p} \quad \text{and} \quad \text{Var}\, X = \frac{1-p}{p^2}.$$

Poisson distribution

A random quantity X is said to be Poisson distributed with parameter λ if

$$P(X = x) = \frac{\lambda^x e^{-\lambda}}{x!}, \; x = 1, 2, \ldots .$$

This distribution is often used for describing the number of events occurring during a specified period of time. The mean and variance of X are both equal to λ.

If X has a binomial distribution with parameters n and p, with n large and p small, the binomial distribution can be accurately approximated by the Poisson distribution with mean np. Consider the occurrence of events X in a time interval $[0, t]$, and divide the interval into a number of small subintervals. Then we may ignore the probability of two or more events occurring in each sub-interval, and the total number of events in $[0, t]$ can be written as a sum of "successes' in a number of Bernoulli trials. It follows that X has a binomial distribution with large n and small p, and can consequently be approximated by a Poisson distribution.

A.1.5 Some Common Continuous Distributions (Models)

Here are some common continuous distributions, again following a standard presentation in the classical setting. Section A.1.6 contains some comments on how to interpret and use these distributions in a framework based on subjective probabilities.

Uniform distribution

A random quantity X is uniformly distributed on the interval (a, b) if it has a probability density given by

$$f(x) = \begin{cases} \dfrac{1}{b-a} & \text{if } a < x < b, \\ 0 & \text{otherwise.} \end{cases}$$

The mean and variance of X are equal to $(b-a)/2$ and $(b-a)^2/12$, respectively.

Exponential distribution

A random quantity X is said to be exponentially distributed with parameter λ (> 0) if

$$P(X \leq x) = 1 - e^{-\lambda x}, \; x \geq 0.$$

Often an exponential lifetime distribution is used for describing the lifetime of a unit, and assume in the following that X represents such a lifetime. For this distribution we have $P(X > u+v \mid X > u) = P(X > v)$, which means that the probability of the unit surviving an additional amount of time v does not depend

BASIC THEORY OF PROBABILITY AND STATISTICS

on how long the unit has functioned. The exponential distribution is the only distribution with this property. This lack of memory simplifies the mathematical modelling.

An important quantity in studying lifetime distributions is the so-called failure rate, $z(x)$, defined by

$$z(x) = \frac{f(x)}{1 - F(x)}, \tag{A.5}$$

where $F(x) = P(X \leq x)$. For the exponential distribution, the failure rate is equal to λ, i.e. independent of time. To see the physical interpretation of the failure rate, consider a small time interval $(x, x + h)$ and assume that the unit has survived x. Then we find that

$$\frac{1}{h} P(X \leq x + h \mid X > x) = \frac{1}{h} \frac{P(x < X \leq x + h)}{P(X > x)}$$

$$= \frac{F(x+h) - F(x)}{h} \frac{1}{1 - F(x)} \to \frac{f(x)}{1 - F(x)} = z(x) \quad \text{when } h \to 0.$$

Thus

$$P(X \leq x + h \mid X > x) \approx z(x) h$$

for small values of h. We see that the failure rate expresses the proneness of the unit to fail at time (age) x. A high failure rate means there is a high probability that the unit will fail soon, whereas a small failure rate means that there is a small probability that the unit will fail in a short time. The cumulative failure rate $\int_0^x z(t) \, dt$ is known as the hazard and is denoted by $Z(x)$.

The mean and variance in the exponential distribution are given by:

$$EX = \frac{1}{\lambda} \quad \text{and} \quad \text{Var } X = \frac{1}{\lambda^2}.$$

Weibull distribution

A random quantity X is said to be Weibull distributed with parameters $\lambda \, (> 0)$ and $\beta \, (> 0)$ if the distribution is given by

$$P(X \leq x) = 1 - e^{-(\lambda x)^\beta}, \quad x \geq 0.$$

We call λ the scale parameter and β the form parameter. If $\beta = 1$ the failure rate becomes a constant. Hence the exponential distribution is a special case of the Weibull distribution. When $\beta > 1$ the failure rate is increasing, and when $\beta < 1$ it is decreasing. Note that

$$1 - F(\lambda^{-1}) = e^{-1} = 0.3679 \quad \text{for all } \beta > 0.$$

The quantity λ^{-1} is often called the *characteristic lifetime*.

The mean (expected) lifetime of the Weibull distribution is given by

$$EX = \frac{1}{\lambda} \Gamma \left(1 + \frac{1}{\beta} \right),$$

where $\Gamma(\cdot)$ is the gamma function defined by

$$\Gamma(x) = \int_0^\infty t^{x-1} e^{-t} \, dt, \quad x > 0.$$

In particular, we have

$$\Gamma(n+1) = n!, \quad n = 0, 1, 2, \ldots.$$

The variance of X becomes

$$\text{Var } X = \frac{1}{\lambda^2}\left[\Gamma\left(1 + \frac{2}{\beta}\right) - \Gamma^2\left(1 + \frac{1}{\beta}\right)\right].$$

Gamma distribution

If $X_1, X_2, \ldots, X_n$ are independent and exponentially distributed random quantities with parameter λ, then $X = X_1 + X_2 + \cdots + X_n$ is gamma distributed with parameters λ and n, i.e.

$$f(x) = \frac{\lambda}{\Gamma(n)} (\lambda x)^{n-1} e^{-\lambda x}, \quad x \geq 0. \tag{A.6}$$

Assume that n units of a certain type have exponentially distributed lifetimes $X_1, X_2, \ldots, X_n$ with failure rate λ and that the units are put into operation one by one as a unit fails. Then the total lifetime equals the sum of the X_i.

The parameter n in (A.6) does not need to be restricted to the positive integers. If it is a positive integer, we can write the survivor function in the following form:

$$1 - F(x) = \sum_{i=0}^{n-1} \frac{(\lambda x)^i}{i!} e^{-\lambda x}.$$

The mean and variance of the gamma distribution are given by

$$EX = \frac{n}{\lambda},$$

$$\text{Var } X = \frac{n}{\lambda^2}.$$

Chi-square distribution

A random quantity X is chi-square distributed with parameter ν if it has a density given by

$$f(x) = \frac{x^{(\nu/2)-1} e^{-x/2}}{2^{\nu/2} \Gamma(\nu/2)}, \quad x \geq 0.$$

The mean of the distribution equals ν and the variance 2ν. The chi-square distribution is closely linked to the gamma distribution. If X has a gamma distribution with parameters (n, λ), then $2\lambda X$ is chi-square distributed with parameter $2n$.

Beta distribution

A random quantity X is said to be beta distributed with parameters a and b if it has a density given by

$$f(x) = \frac{\Gamma(a+b)}{\Gamma(a)\Gamma(b)} x^{a-1}(1-x)^{b-1},$$

for $x \geq 0$, and $a > 0$, $b > 0$. The mean and variance are equal to $a/(a+b)$ and $ab/[(a+b)^2(a+b+1)]$, respectively.

Beta-binomial distribution

A random quantity X is said to be beta-binomial distributed with parameters (n, a, b) if it has a density given by

$$f(x) = \binom{n}{x} \frac{\Gamma(a+x)\Gamma(n+b-x)\Gamma(a+b)}{\Gamma(n+a+b)\Gamma(a)\Gamma(b)},$$

for $x = 0, 1, 2, \ldots, n$, $a > 0$, $b > 0$ and $n = 0, 1, 2, \ldots$. The mean and variance are equal to $na/(a+b)$ and $nb(n+a+b)/[(a+b)^2(a+b+1)]$, respectively.

Triangular distribution

A random quantity X is triangle distributed with parameters a, b and c if it has a density given by

$$f(x) = \begin{cases} \dfrac{2(x-a)}{(b-a)(c-a)} & \text{if } a \leq x \leq b, \\ \dfrac{2(c-x)}{(c-a)(c-b)} & b < x \leq c. \end{cases}$$

The density increases linearly from a to b, and then decreases linearly from b to c. The mean and variance are equal to $(a+b+c)/3$ and $(a^2+b^2+c^2-ab-ac-bc)/18$, respectively.

Normal distribution

The random quantity X is said to be normally distributed with parameters μ and σ^2 if it has a density given by

$$f(x) = \frac{1}{\sqrt{2\pi}} \exp\left\{-\left(\frac{x-\mu}{\sigma}\right)^2\right\}, \quad -\infty < x < \infty.$$

It can be shown that $EX = \mu$ and $\text{Var } X = \sigma^2$. If $\mu = 1$ and $\sigma = 1$, the distribution is called a standard normal distribution. The normal distribution is probably the most widely used distribution in the entire field of statistics and probability. It turns out that the means of a number of populations exhibit a bell-shaped (i.e. normal) curve. The central limit theorem gives a precise mathematical formulation of this fact.

Theorem A.2 *Let $X_1, X_2, \ldots$, be a sequence of independent random quantities having a common distribution, and let $EX_i = \mu$ and $\text{Var } X_i = \sigma^2$. Then,*

$$\frac{X_1 + X_2 + \cdots + X_n - n\mu}{\sigma\sqrt{n}}$$

tends to the standard normal distribution with mean 0 and variance 1, i.e.

$$P\left(\frac{X_1 + X_2 + \cdots + X_n - n\mu}{\sigma\sqrt{n}} \leq a\right) \to \frac{1}{\sqrt{2\pi}} \int_{-\infty}^{a} e^{-x^2/2} \, dx,$$

as $n \to \infty$.

Lognormal distribution

A random quantity X is said to be lognormally distributed with parameters μ and σ^2 if $\ln X$ has a normal distribution with parameters μ and σ^2.

Multivariate normal distribution

Let $Z_1, Z_2, \ldots, Z_n$ be a set of n independent standard normally distributed random quantities. If for some constants a_{ij} and μ_i, we can write

$$X_i = \mu_i + \sum_{j=1}^{n} a_{ij} Z_j,$$

for $i = 1, 2, \ldots, m$, then the random quantities $X_1, X_2, \ldots, X_m$ have a multivariate normal distribution. This distribution is completely specified by the knowledge of the values of all EX_i and $\text{Cov}(X_i, X_j)$. It can be shown that any linear combination of the X_i is a normally distributed random quantity.

If n is equal to 2, the multivariate normal distribution is known as the bivariate normal distribution.

A.1.6 Some Remarks on Probability Models and Their Parameters

The above review of commonly used distribution classes is in accordance with a classical view of probability. The random quantities (variables) have a true, underlying distribution and the distribution class is a model of this distribution; that is, it is a representation of the real world. By statistical inference we seek to identify the best parameter value, in the sense that it gives the most accurate representation of the world.

In a framework based on subjective probabilities, called the Bayesian framework or the Bayesian approach, it is not obvious how to interpret and use probability distribution classes or models. Can we speak about models in this case? Well, according to the Bayesian approach, all probabilities are subjective probabilities, based on judgements, reflecting our uncertainty about something.

BASIC THEORY OF PROBABILITY AND STATISTICS

Probabilities are always conditioned on the background information, K say. To specify the probabilities related to a random quantity X, a direct assignment could be used, based on everything we know. Since this knowledge is often complex, of high dimension, and much in K may be irrelevant to X, this approach is often replaced by probability models, which is a way of abridging K to make it manageable. Such probability models play a key role in the Bayesian approach. A probability model, $p(x|\theta)$, expresses the probability distribution of the unknown quantity X given a parameter θ. This parameter θ is unknown, it is a random quantity and our uncertainty related to its value is specified through a prior distribution $P(\theta)$. According to the law of total probability,

$$P(X \leq x) = \int p(x|\theta)\,\mathrm{d}P(\theta). \tag{A.7}$$

More precisely, showing the dependence on the background information K,

$$P(X \leq x|K) = \int P(X \leq x|\theta, K)\,\mathrm{d}P(\theta|K). \tag{A.8}$$

If we knew θ, we would judge X independent of K, so that for all θ, $P(X \leq x|\theta, K) = P(X \leq x|\theta)$, then equation (A.8) is equal to (A.7). Thus the uncertainty distribution of X is expressed via two probability distributions, $p(x|\theta)$ and $P(\theta|K)$. The latter distribution is the prior distribution of θ. The two distributions reflect what is commonly known as aleatory (stochastic) uncertainty and epistemic (state of knowledge) uncertainty. If more data become available, the prior distribution is updated to the posterior distribution using Bayes theorem. See Section 2.3.4 page 37 and Section 4.3.4 page 79 for a discussion of this interpretation.

A.1.7 Random Processes

A *random process* (*stochastic process*) $X(t)$, $t \in T$, is a collection of random quantities. That is, for each $t \in T$, $X(t)$ is a random quantity. The index t is often interpreted as time and, as a result, we refer to $X(t)$ as the state of the process at time t. The set T is called the *index set* of the process. In this book T is usually $[0, \infty)$ or $\{0, 1, 2, \ldots\}$. We shall look at just one example of random processes, the Poisson process.

The Poisson process

Consider a sequence of events occurring at times $S_1, S_2, \ldots$, and let T_i denote the interarrival times given by $T_i = S_i - S_{i-1}$, $i = 1, 2, \ldots$, where $S_0 = 0$. Furthermore let $N(t)$ denote the number of events that have occurred before or at time t, i.e.

$$N(t) = \max\{i : S_i \leq t\}.$$

The random process $N(t)$ is called a counting process. If the random quantities T_i are independent and identically distributed, the process is called a renewal process and if in addition the lifetime distribution is exponential with parameter

λ, $N(t)$ is a Poisson process with parameter λ. It can be shown that if $N(t)$ is a Poisson process with parameter λ, then $N(t)$ is Poisson distributed with parameter λt, i.e.

$$P(N(t) = i) = \frac{(\lambda t)^i}{i!} e^{-\lambda t}, \quad i = 0, 1, 2, \ldots.$$

Thus

$$EN(t) = \lambda t.$$

For a counting process we define the associated intensity process $\lambda(t)$ by

$$\lim_{h \to 0} \frac{E[N(t+h) - N(t) | N(u), u \leq t]}{h} = \lambda(t). \quad (A.9)$$

In the Poisson process case $\lambda(t)$ is equal to the constant λ, i.e. the intensity does not depend on the history of the process up to time t. Thus, if the expected number of failures per unit of time is independent of the history and is equal to a constant λ, the process is a Poisson process with rate λ.

A.2 CLASSICAL STATISTICAL INFERENCE

This section reviews some elementary statistical inference in a classical context. The emphasis is on estimation.

A.2.1 Non-Parametric Estimation

Consider a random variable X, having probability distribution $F(x) = P(X \leq x)$. The task is to estimate this distribution given observations $X_1, X_2, \ldots, X_n$. The random variable X has a distribution function F. All the random variables are assumed independent.

Often the data are censored, i.e. we do not observe X_i, but $\min\{X_i, C_i\}$, where C_i is the censoring time. We will, however, not discuss this case any further here.

As an estimator for $F(x)$ we may use the empirical distribution function, $\widehat{F}(x)$, defined by

$$\widehat{F}(x) = \frac{1}{n} \sum_i I(X_i \leq x),$$

where I is the indicator function, which equals 1 if the argument is true and 0 otherwise. If $n \to \infty$, then $\widehat{F}(x) \to F(x)$ with probability one.

For non-negative observations the Nelson–Aalen estimator is often used. This is an estimator of the cumulative failure rate $Z(x) = \int_0^x z(t)\,dt$, cf. (A.5), and is given by

$$Z^*(t) = \sum_{\{i :, X_i \leq t\}} \frac{1}{n - i + 1}.$$

Based on estimators as above we can make plots and fit the distribution to a parametric class of distributions, like the exponential distribution. If we compute the Nelson–Aalen estimator and the plot is close to a straight line starting at the origin, this would indicate that an exponential distribution may be appropriate as the hazard of this distribution is such a straight line.

In this framework a particular probability model can be formally evaluated via so-called 'goodness-of-fit' tests. The idea is to use a measure of distance between the empirical distribution and the underlying theoretical distribution. Consult textbooks in statistics for the details.

A.2.2 Estimation of Distribution Parameters

We assume that the distribution $F(x)$ belongs to a known parametric class of distributions, for example the exponential class or the normal class. The problem is to estimate the parameters of the distribution. As above we assume that we have observations $X_1, X_2, \ldots, X_n$.

Maximum likelihood estimation

We begin with the Poisson distribution. Let

$$f(x|\lambda) = \frac{\lambda^x e^{-\lambda}}{x!}.$$

The probability distribution related to the observed data $X_i = x_i$ then becomes

$$\prod_{i=1}^{n} f(x_i|\lambda) = \lambda^{x_1+x_2+\cdots+x_n} e^{-n\lambda} \prod_{i=1}^{n} \frac{1}{x_i!}.$$

As a function of the parameter λ, this probability is called the likelihood function and is denoted $L(\lambda)$. The likelihood function is a measure of the likelihood of the observed result as a function of the unknown parameter. The maximum likelihood estimate (MLE) of λ is denoted λ^* and it maximizes $L(\lambda)$. In other words, MLE is the value of λ that makes the observed result most likely. In practice the MLE is determined by differentiating the likelihood function and setting the derivative equal to zero. By doing this, we obtain

$$\lambda^* = \frac{x_1 + x_2 + \cdots + x_n}{n},$$

i.e. the average number of the observations.

Now consider any distribution and let $\theta = (\theta_1, \theta_2, \ldots, \theta_p)$ be the parameter of the distribution. Furthermore, let $f(t|\theta)$ denote the probability density in the continuous case and the distribution $P(X = x)$ in the discrete case. For the exponential distribution $\theta = \lambda$, and for the normal distribution $\theta = (\mu, \sigma^2)$. In this set-up the likelihood function is given by

$$L(\theta) = \prod_{i=1}^{n} f(x_i|\theta).$$

For the exponential distribution with $\theta = \lambda$, we find that

$$\lambda^* = \frac{n}{x_1 + x_2 + \cdots + x_n},$$

and for the normal distribution we find that

$$\mu^* = \frac{x_1 + x_2 + \cdots + x_n}{n},$$

$$(\sigma^2)^* = \frac{1}{n-1} \sum_{i=1}^{n} (x_i - \mu^*)^2.$$

In general it is not possible to find an explicit expression for the MLE. Numerical methods must then be used.

Confidence interval

As a measure of data variation, a confidence interval (region) for θ is often presented in addition to the estimate of the parameter.

An interval (θ_L, θ_H) is said to be a $(1-\alpha)100\%$ confidence interval if there is a probability of $1-\alpha$ that the interval contains θ, that is

$$P(\theta_L < \theta < \theta_H) = 1 - \alpha.$$

The level $1-\alpha$ is a measures of our confidence that the interval contains θ. The most common values of α are 0.10 and 0.05. Notice that θ_L and θ_H are random variables. When the confidence interval is calculated, i.e. we observe specific values of θ_L and θ_H, the resulting interval either contains the true value of θ or it does not, but in the long run if the experiment were repeated many times, then θ would be included in the confidence interval $(1-\alpha)100\%$ of the times. The level of confidence $1-\alpha$ therefore expresses a property of the method that we are using to determine the interval.

In the exponential model a $(1-\alpha)100\%$ confidence interval is given by

$$(\lambda_L, \lambda_H) = \left(\frac{z_{\alpha/2, 2n}}{2(X_1 + X_2 + \cdots + X_n)}, \frac{z_{1-\alpha/2, 2n}}{2(X_1 + X_2 + \cdots + X_n)} \right)$$

where $z_{\alpha, v}$ equals the α quantile in the chi-square distribution with v degrees of freedom. The α quantile of the distribution of a random variable X is the value x_α such that $P(X \leq x_\alpha) = \alpha$. For the Poisson process model with rate λ and observed in an interval $[0, t]$, we can use the following interval:

$$(\lambda_L, \lambda_H) = \left(\frac{z_{\alpha/2, 2N}}{2t}, \frac{z_{1-\alpha/2, 2(N+1)}}{2t} \right),$$

where N is the number of events observed in $[0, t]$. Consider the following example.

Example A.3 *We consider the problem of specifying a 90% confidence interval for the rate λ in a Poisson process, based on the data given in Section 2.1.2. During 12 years of observation, 12 leakages are registered. This gives an estimated* rate of 1 per year. A 90% confidence interval, given these data, is

$$\lambda_L = \frac{13.85}{2 \times 12} = 0.58$$

$$\lambda_H = \frac{38.89}{2 \times 12} = 1.62.$$

A.2.3 Testing Hypotheses

The set-up is as above. We assume that the distribution $F(x)$ belongs to a known parametric class of distributions and that we have available observations $X_1, X_2, \ldots, X_n$. We use the binomial model with parameters n and p to illustrate ideas. The observation X_i here refers to 'success' in the ith experiment, such that the sum of the X_is is the total number of observed 'successes'. This sum is prior observation seen as a random variable, and we denote it by Y.

The problem is now to formulate a statistical test. We do this by formulating statements about the parameter of the probability model; in this case the success probability p. The starting point is the null hypothesis, H_0, which we may think of as '$p = 0.25$', say. The test questions the truth of this statement in relation to an alternative hypothesis H_1, say '$p > 0.25$'. If the data provide sufficient support, we assert that H_0 is false and H_1 is correct. We conclude in this way if we have a high confidence about the correctness of H_1. As a concrete example, consider a medical treatment that is known to have a 'success' rate of 25%. An adjustment of this treatment is considered, and the question is whether this adjustment would increase the 'success' rate. It is reasonable to assert that $p > 0.25$ if the number of successes is large, i.e. $Y \geq k$, for a suitable choice of k. We see that if k is specified, the test is specified. Let α be the probability that $Y \geq k$ if H_0 is true, i.e. $p = 0.25$. These probabilities for various k are found from statistical tables for the binomial distribution, or use of approximations to the normal distribution.

We search for a k such that α becomes rather small, say 0.05 or 0.10. For example, if $n = 20$ and $\alpha = 0.10$, we find that $k = 8$, which corresponds to a fraction of successes of 40%. If we observe 8 or more successes, the result is so 'extreme' relative to H_0, that we reject H_0. We refer to α as the significance level of the test. It is the probability of an error of type I, i.e. of rejecting H_0 when in fact it is true. It should be rather low as it represents a probability of making a wrong conclusion—asserting H_1 if H_0 is true. On the other hand, specifying a very low value of α means that the probability of not concluding that H_1 is true if it is in fact true, becomes high. So a balance has to be achieved. The probability of this latter type of error is denoted β, and it is a function of the parameter value. This type of error is called an error of type II. In our example, if $p = 0.30$, the probability that we do not reject H_0, the type II error probability $P(Y < 8|p = 0.3)$ is about 77%. We see that to reject H_0 a rather extreme observation is required using the above principles. The point is that type I errors are considered more serious than type II errors. In the medical treatment example, the starting point is that there is no improvement. Only if the data give very strong support for the alternative hypothesis, should we reject

H_0; the probability of a failure of type I should be small. Note that when not rejecting H_0, we do not say that H_0 is true; the conclusion is that we do not have statistical evidence to reject the null hypothesis.

A.2.4 Regression

Regression analysis is mainly used for prediction. By developing a statistical model, the values of a dependent or response variable Y is predicted based on the values of an independent variable X. As an example, an economist might want to develop a statistical model that predicts how much money a population of people would spend (Y) based on how much money they earn (X). The simplest type of regression analysis is based on a linear regression model. To develop the model, we assume that a sample of n independent observations $(X_1, Y_1), (X_2, Y_2), \ldots, (X_n, Y_n)$ is obtained, where X_i represents the ith value of the independent variable X and where Y_i represents the corresponding response; that is, the ith value of the dependent variable Y. The linear regression model specifies that there is an underlying true relationship between EY and EX, expressed by a linear function. In practice this linear function is not realized because of randomness. Mathematically, these ideas are formulated as

$$Y_i = \beta_0 + \beta_1 X_i + \epsilon_i,$$

where ϵ_i is the random error in Y for observation i, and β_0 and β_1 are constants to be estimated. We see that β_1 represents the slope of the line $Y = \beta_0 + \beta_1 X$ and β_0 represents the intercept of the line with the Y-axis. We may think of this underlying straight line as a model of the true relationship between EY and EX for a large (infinite) population of which the sample of n belongs to. The random variables ϵ_i represent the error terms. A common model for these error terms is the normal distribution with mean zero and variance σ^2. This distribution reflects the variations of the observations Y around their expected values.

To estimate the parameters β_i, the standard technique is to apply the method of least squares, i.e. to identify the values that minimize the sum of squared errors in the sample. We denote the estimators β_i^* and they are given by

$$\beta_0^* = \overline{Y} - \beta_1^* \overline{X},$$

$$\beta_1^* = \frac{\sum_{i=1}^n (X_i - \overline{X}) Y_i}{\sum_{i=1}^n (X_i - \overline{X})^2},$$

where $\overline{X} = \sum_{i=1}^n X_i / n$ and $\overline{Y} = \sum_{i=1}^n Y_i / n$. To predict Y based on X we use the line $Y = \beta_0^* + \beta_1^* X$.

To estimate the variance σ^2, the common estimator is

$$S^2 = \frac{1}{n-2} \sum_{i=1}^n (Y_i - \beta_0^* + \beta_1^* X_i)^2.$$

Confidence intervals and statistical tests can now be derived for the parameters β_i and σ^2. We refer to textbooks in statistics.

A.3 BAYESIAN INFERENCE

To illustrate the Bayesian approach to statistical inference, we first consider the Poisson distribution. Suppose we are interested in the number X of failures of a system in operation during a specific period of time in the future. We assume that X is Poisson distributed with parameter $\lambda = 1$ or $\lambda = 2$. We may think of these two parameter values as corresponding to two alternative types of system, type 1 and 2. Now, suppose that we have just one observation, x_1. The MLE of λ is then equal to x_1. Following the Bayesian approach, we also include the analysts' knowledge (uncertainty) related to the value of λ, before observing x_1. This knowledge is expressed by subjective probabilities P_1 and P_2, with $P_1 + P_2 = 1$. We call P_i the prior distribution of λ. Given the observation $X_1 = x_1$, we obtain the posterior distribution P_i^D reflecting our knowledge (uncertainty) about the value of λ given the data:

$$P_i^D = cf(x_1|i)P_i, \quad i = 1, 2, \tag{A.10}$$

where c is a constant such that $P_1^D + P_2^D = 1$ and $f(x|\lambda) = \lambda^x e^{-\lambda}/x!$. To establish this posterior distribution we have used Bayes' formula, which gives

$$P_\lambda^D = P(\lambda|X_1 = x_1) = \frac{P(X_1 = x_1|\lambda)P_\lambda}{P(X_1 = x_1)} - cf(x_1|\lambda)P_\lambda, \ \lambda = 1, 2.$$

Suppose that $P_i = 0.5$ and that we have observed $X_1 = 1$. Then (A.10) gives a posterior distribution $P_1^D = 0.58$ and $P_2^D = 0.42$. It is natural to estimate λ by 1 as $P_1^D > P_2^D$. In this situation we can divide the uncertainty into the stochastic (aleatory) uncertainty given by the Poisson distribution and the state-of-knowledge (epistemic) uncertainty related to the true value of λ expressed by the posterior distribution. As long as we stick to the Poisson distribution, additional information will change only the state-of-knowledge uncertainty distribution.

Using the law of total probability, we can establish the so-called predictive distribution of X:

$$P(X = x) = f(x|1)P_1^D + f(x|2)P_2^D. \tag{A.11}$$

This distribution reflects both the stochastic and the state-of-knowledge uncertainty. Inserting the numerical values in (A.11), we find that $P(X = 0) = 0.27$.

Now let us return to the general setting in Section A.2.2, with θ as the unknown parameter. If $p(\theta)$ expresses the a prior probability density, then the posterior density $p(\theta \mid Data)$, is given by

$$p(\theta \mid Data) = c L(\theta) p(\theta), \tag{A.12}$$

where c is a constant such that the integral over $p(\theta \mid Data)$ is 1. The posterior density expresses our uncertainty with respect to the true value of θ when the data are observed, and includes all the available information about θ. Based on the posterior distribution, we can establish estimators and *credibility intervals*. This will be illustrated below for the exponential distribution.

Assume in the rest of this section that the underlying lifetime distribution is exponential with failure rate λ. If the a priori density $p(\lambda)$ takes the form

$$p(\lambda) = b^a \lambda^{a-1} e^{-b\lambda} / \Gamma(a),$$

i.e. p is a gamma density with parameters a and b, then we find that the posterior density is also a gamma density, with parameters $a + n$ and $b + y$, where $y = x_1 + x_2 + \cdots + x_n$. This means that the exponential and the gamma distributions are *conjugate*—the distribution classes of the prior and the posterior are the same. Writing $F(x|\lambda) = 1 - e^{-\lambda x}$, the predictive distribution is given by

$$P(X_i \leq x) = E[F(x|\lambda)] = \int (1 - e^{-\lambda x}) p^D(\lambda) \, d\lambda, \qquad (A.13)$$

where $p^D(\lambda)$ is the posterior density function of λ. A natural estimator for λ is the mean of the posterior distribution

$$\hat{\lambda} = \frac{a+n}{b+y}. \qquad (A.14)$$

A $(1-\alpha)100\%$ credibility interval for λ, (λ_L, λ_H), is determined by the posterior probability

$$P(\lambda_L < \lambda < \lambda_H \mid Data) = 1 - \alpha.$$

With a gamma a priori density it can be shown that the interval

$$\left(\frac{z_{\alpha/2, 2(a+n)}}{2(b+y)}, \frac{z_{1-\alpha/2, 2(a+n)}}{2(b+y)} \right)$$

is a $(1-\alpha)100\%$ credibility interval, where $z_{\alpha, v}$ equals the α quantile in the chi-square distribution with v degrees of freedom. Note that a credibility interval is interpreted given that the data are observed, in contrast to the classic confidence interval, which is interpreted before the data are observed.

A critical part of the Bayesian analysis is to establish the a priori distribution. The choice of a gamma a priori density gives simple mathematics. In addition it can be shown that if the a priori distribution is non-informative, i.e.

$$p(\lambda) = \begin{cases} \dfrac{1}{M} & \text{for } 0 \leq \lambda \leq M \ (M \text{ large}), \\ 0 & \text{otherwise}, \end{cases}$$

then the resulting posterior distribution is an approximate gamma distribution. We can think of a situation where we started with a non-informative a priori distribution and this distribution was updated to a gamma distribution. The parameter $a - 1$ in the gamma distribution can be interpreted as the number of observations in an earlier experiment (real or fictional) and b as the corresponding test time. See Section 4.3.4 page 82 for a more detailed discussion on the specification of prior distributions.

In the Bayesian theory, it is often referred to as the likelihood principle. It states that the only contribution the data make to inference is through the likelihood function for the observed data. This principle renders significance tests not acceptable. The likelihood therefore plays a more important role in Bayesian statistics than it does in the frequentist form, yet the likelihood alone is not adequate for inference but needs to be tempered by the parameter distribution, see (A.12).

If data are available to compare with the predictive distributions, we can obtain an assessment of the predictive ability of a proposed model. The purpose of this type of assessment is to evaluate the 'goodness' of the models as a basis for selecting a proper model. In the Bayesian paradigm, model selection is formally done via Bayes factors and prequential prediction; see Singpurwalla and Wilson (1999).

A.3.1 Statistical (Bayesian) Decision Analysis

We briefly review the basic formalism of statistical decision analysis. A decision-maker has to choose a single action a from a space of possible actions A. Features of the world are modelled by an unknown state of nature θ, which is known to lie in a set of possible states of nature Θ. If the decision-maker chooses action a and θ is the state of nature, the consequence equals $c(a, \theta)$, which is possibly multidimensional or multi-attributed. Before choosing the action, the decision-maker may observe an outcome $X = x$ of an experiment, which depends on the unknown state θ. The distribution of X is denoted by $p(x|\theta)$. The decision-maker's objectives are encoded in a real-valued loss function $l(a, \theta)$, which measures the loss or negative utility of the consequence $c(a, \theta)$, i.e. in everyday terms, it measures the value or worth of the consequence to the decision-maker. The problem for the decision-maker is to choose an action $d(x)$ to minimize in some sense $l(d(x), \theta)$. The notation $d(x)$ emphasizes we are seeking to identify a decision rule that suggests which action to take when $X = x$ has been observed. Since θ is unknown, this is not straightforward and several approaches have been suggested. Two of them are the minimax approach and the Bayesian approach.

The minimax solution is to define $d(\cdot)$ by minimizing over the set of all possible decision rules the maximum expected loss with respect to θ, where the expectation is taken with respect to $p(x|\theta)$, i.e. choose $d(\cdot)$ such that

$$\max_{\theta} E\left[l(d(X), \theta)|\theta\right]$$

is minimized.

The Bayesian solution is to encode the decision-maker's prior knowledge of θ through the prior distribution $p(\cdot)$. The knowledge is updated through the use of Bayes' theorem to obtain the posterior distribution $p(\theta|x)$. Then the decision-maker should choose the action $d(x) = a$ to minimize their posterior expected loss, i.e. choose a such that

$$E[l(a, \theta)|x]$$

is minimized, where expectation is with respect to θ.

This decision framework can also be used for classical statistical inference.

BIBLIOGRAPHIC NOTES

Some references providing more comprehensive and detailed overviews of probability theory and statistical inference are Bedford and Cooke (2001), Lindley (2000), Ross (1993), Singpurwalla and Wilson (1999) and Vose (2000). Confidence intervals for the exponential and Poisson distributions are given in Bain and Engelhardt (1991). The review of statistical decision theory is based on French and Insua (2000).

Appendix B

Terminology

This appendix summarizes some risk analysis and management terminology used in the book. Unless stated otherwise, the terminology is in line with the standard developed by the ISO TMB Working Group on risk management terminology (ISO 2002). ISO is the International Organization for Standardization.

The relationships between the terms and definitions for risk management are shown following the definitions. Risk management is part of the broader management processes of organizations.

1. **aleatory uncertainty**
   variation of quantities in a population
   This definition is not given in the ISO standard.
2. **consequence**
   outcome of an **event**
   There may be one or more consequences from an event. Consequences may range from positive to negative. Consequences may be expressed qualitatively or quantitatively.
3. **epistemic uncertainty**
   lack of knowledge about the 'world' (i.e. the system performance), and **observable quantities** *in particular*
   In our framework, uncertainty is the same as epistemic uncertainty. In a classical approach to risk analysis, epistemic uncertainty means uncertainty about the (true) value of a parameter of a probability model.
   This definition is not given in the ISO standard.
4. **event**
   occurrence of a particular set of circumstances
5. **interested party**
   person or group having an interest in the performance of an organization
   Examples are customers, owners, employees, suppliers, bankers, unions, partners or society.

A group may be an organization, part of an organization, or more than one organization.

6. **mitigation**
   limitation of any negative consequence of a particular event

7. **observable quantity**
   quantity expressing a state of the 'world', i.e. a quantity of the physical reality or nature, that is unknown at the time of the analysis but will, if the system being analysed is actually implemented, take some value in the future, and possibly become known
   This definition is not given in the ISO standard.

8. **probability**
   a measure of uncertainty of an event
   This definition can be seen as a special case of the definition given by the ISO standard: 'extent to which an event is likely to occur'.

9. **residual risk**
   the **risk** *remaining after* **risk treatment**

10. **risk**
    uncertainty of the performance of a system (the world), quantified by **probabilities** *of observable quantities*
    When risk is quantified in a risk analysis, this definition is in line with the ISO standard definition: 'combination of the **probability** of an **event** and its **consequence**'.

11. **risk acceptance**
    a decision to accept a **risk**
    Risk acceptance depends on **risk criteria**

12. **risk acceptance criterion**
    a reference by which risk is assessed to be acceptable or unacceptable
    This definition is not included in the ISO standard. It is an example of a risk criterion.

13. **risk analysis**
    systematic use of information to identify **sources** *and assign* **risk** *values*
    Risk analysis provides a basis for **risk evaluation, risk treatment** and **risk acceptance**. Information can include historical data, theoretical analysis, informed opinions, and concerns of **stakeholders**.

14. **risk assessment**
    overall process of **risk analysis** *and* **risk evaluation**

15. **risk avoidance**
    decision not to become involved in, or action to withdraw from a risk situation
    The decision may be taken based on the result of **risk evaluation**.

16. **risk communication**
    exchange or sharing of information about **risk** *between the decision-maker and other* **stakeholders**
    The information may relate to the existence, nature, form, **probability**, severity, acceptability, treatment or other aspects of **risk**.

17. **risk control**
    actions implementing **risk management** *decisions*

Risk control may involve monitoring, re-evaluation, and compliance with decisions.

18. **risk criteria**
terms of reference by which the significance of **risk** *is assessed*
Risk criteria may include associated cost and benefits, legal and statutory requirements, socio-economic and environmental aspects, concerns of **stakeholders**, priorities and other inputs to the assessment.

19. **risk evaluation**
process of comparing risk against given **risk criteria** *to determine the significance of the* **risk**
Risk evaluation may be used to assist the decision-making process.

20. **risk financing**
provision of funds to meet the cost of implementing **risk treatment** *and related costs*

21. **risk identification**
process to find, list and characterize elements of **risk**
Elements may include **source**, **event**, **consequence**, **probability**. Risk identification may also identify **stakeholder** concerns.

22. **risk management**
coordinated activities to direct and control an organization with regard to **risk**
Risk management typically includes **risk assessment**, **risk treatment**, **risk acceptance** and **risk communication**.

23. **risk management system**
set of elements of an organization's management system concerned with managing **risk**
Management system elements may include strategic planning, decision-making, and other processes for dealing with **risk**

24. **risk optimization**
process, related to a **risk**, *to minimize the negative and to maximize the positive* **consequences** *and their respective* **probabilities**
In a safety context **risk optimization** is focused on reducing the **risk**.

25. **risk perception**
set of values or concerns with which a **stakeholder** *views* **risk**
Risk perception depends on the **stakeholder's** needs, issues and knowledge.

26. **risk quantification**
process used to assign values to **risk**
In the ISO standard on risk management terminology, the term 'risk estimation' is used, with the definition 'process used to assign values to the **probability** and **consequence** of a risk'.

27. **risk reduction**
actions taken to reduce **risk**
This definition extends the ISO standard definition: 'actions taken to lessen the **probability**, negative **consequences**, or both, associated with a **risk**'.

28. **risk retention**
acceptance of the burden of loss or benefit of gain from a **risk**

Risk retention includes the acceptance of **risks** that have not been identified. **Risk retention** does not include treatments involving insurance, or transfer by other means.

29. risk transfer
    share with another party the benefit of gain or burden of loss for a **risk**
    Risk transfer may be effected through insurance or other agreements. **Risk transfer** may create new **risks** or modify existing **risk**. Legal or statutory requirements may limit, prohibit or mandate the transfer of certain **risk**.

30. risk treatment
    process of selection and implementation of measures to modify **risk**
    The term '**risk treatment**' is sometimes used for the measures themselves. **Risk treatment** measures may include avoiding, optimizing, transferring or retaining risk.

31. source
    thing or activity with a potential for **consequence**
    Source in a safety context is a hazard.

32. source identification
    process to find, list and characterize **sources**
    In the safety literature, **source identification** is called hazard identification.

33. stakeholder
    any individual, group or organization that may affect, be affected by, or perceive itself to be affected by the **risk**
    The decision-maker is also a **stakeholder**. The term 'stakeholder' includes, but has a broader meaning than '**interested party**'.

34. uncertainty
    lack of knowledge about the performance of a system (the 'world'), and **observable quantities** *in particular*
    This definition is not given in the ISO standard.

RISK MANAGEMENT: RELATIONSHIPS BETWEEN KEY TERMS

- **Risk assessment**
  - Risk analysis
    - *Source identification*
    - *Risk quantification*
  - Risk evaluation
- **Risk treatment**
  - Risk avoidance
  - Risk optimization
  - Risk transfer
  - Risk retention
- Risk acceptance
- Risk communication

Bibliography

Ale, B. (1999) Trustnet: finding new ways to deal with risks. *ESRA Newsletter*, April.

Allison, G. and Zelikow, P. (1999) *Essence of Decision – Explaining the Cuban Crisis*, 2nd edn, Addison Wesley Longman.

Ang, A.S. and Tang, W.H. (1984) *Probability Concepts in Engineering Planning and Design*, John Wiley & Sons, Inc., New York.

Apeland, S. and Aven, T. (2000) Risk based maintenance optimization: foundational issues. *Reliability Engineering and System Safety*, **67**: 285–292.

Apeland, S., Aven, T. and Nilsen, T. (2002) Quantifying uncertainty under a predictive epistemic approach to risk analysis. *Reliability Engineering and System Safety*, **75**: 93–102.

Apostolakis, G. (ed.) (1988) *Reliability Engineering and System Safety*, vol. 23, no. 4.

Apostolakis, G. (1990) The concept of probability in safety assessments of technological systems. *Science*, **250**: 1359–1364.

Apostolakis, G. and Mosleh, A. (1986) The assessment of probability distributions from expert opinions with an application to seismic fragility curves. *Risk Analysis*, **6**: 447–461.

Apostolakis, G. and Wu, J.S. (1993) The interpretation of probability, De Finetti's representation theorem, and their implications to the use of expert opinions in safety assessment. In *Reliability and Decision Making*, Barlow, R.E. and Clarotti, C.A. (eds), Chapman & Hall, London, pp. 311–322.

Armstrong, J.S. (1985) *Long-Range Forecasting: From Crystal Ball to Computer*, John Wiley & Sons, Inc., New York.

Arrow, K.J. (1951) *Social Choice and Individual Values*, John Wiley & Sons, Inc., New York.

Aven, T. (1992) *Reliability and Risk Analysis*, Elsevier, London.

Aven, T. (2000a) Risk analysis – a tool for expressing and communicating uncertainty. In *Proceedings of the European Safety and Reliability Conference*, pp. 21–28.

Aven, T. (2000b) Reliability analysis as a tool for expressing and communicating uncertainty. In *Recent Advances in Reliability Theory: Methodology, Practice and Inference*, Birkhäuser, Boston, pp. 23–28.

Aven, T. (2001) On the practical implementation of the Bayesian paradigm in reliability and risk analysis. In *System and Bayesian Reliability: Essays in Honor of Professor Richard E. Barlow*, Hayakawa, Y. and Xie, M. (eds), World Scientific, London, pp. 269–286.

Aven, T. and Jensen, U. (1999) *Stochastic Models in Reliability*, Springer-Verlag, New York.
Aven, T. and Kørte, J. (2003) On the use of cost/benefit analyses and expected utility theory. *Reliability Engineering and System Safety*, **79**: 289–299.
Aven, T. and Kvaløy, J.T. (2002) Implementing the Bayesian paradigm in practice. *Reliability Engineering and System Safety*, **78**: 195–201.
Aven, T. and Pitblado, R. (1998) On risk assessment in the petroleum activities on the Norwegian and the UK continental shelves. *Reliability Engineering and System Safety*, **61**: 21–30.
Aven, T. and Pörn, K. (1998) Expressing and interpreting the results of quantitative risk analyses. Review and discussion. *Reliability Engineering and System Safety*, **61**: 3–10.
Aven, T. and Rettedal, W. (1998) Bayesian frameworks for integrating QRA and SRA. *Structural Safety*, **20**: 155–165.
Aven, T., Nilsen, E.F. and Nilsen, T. (2003) Expressing economic risk – review and presentation of a unifying approach. *Risk Analysis*, forthcoming.
Bain, L.J. and Engelhardt, M. (1991) *Statistical Analysis of Reliability and Life-testing Models*, Marcel Dekker, New York.
Barlow, R.E. (1998) *Engineering Reliability*, SIAM, Philadelphia PA.
Barlow, R.E. and Clarotti, C.A. (1993) *Reliability and Decision Making*, Preface, Chapman & Hall, London.
Barlow, R.E. and Proschan, F. (1975) *Statistical Theory of Reliability and Life Testing*, Holt, Rinehart and Winston, New York.
Beck, U. (1992) *Risk Society*, Sage, London.
Bedford, T. and Cooke, R. (1999) A new generic model for applying MAUT. *European Journal of Operational Research*, **118**: 589–604.
Bedford, T. and Cooke, R. (2001) *Probabilistic Risk Analysis*, Cambridge University Press, Cambridge.
Bell, D.E., Raiffa, H. and Tversky, A. (eds) (1988) *Decision Making*, Cambridge University Press, Cambridge.
Berg Andersen, L., Nilsen, T., Aven, T. and Guerneri, A. (1997) A practical case of assessing subjective probabilities – a discussion of concepts and evaluation of methods. In *Proceedings of the European Safety and Reliability Conference*, pp. 209–216.
Bernardo, J. and Smith, A. (1994) *Bayesian Theory*, John Wiley & Sons, Inc., New York.
Bernstein, P. (1996) *Against the Gods*, John Wiley & Sons, Inc., New York.
Blockley, D. (ed.) (1992) *Engineering Safety*, McGraw-Hill, New York.
Clemen, R.T. (1996) *Making Hard Decisions*, 2nd edn, Duxbury Press, New York.
Cooke, R.M. (1991) *Experts in Uncertainty: Opinion and Subjective Probability in Science*, Oxford University Press, New York.
Copeland, T.E. and Weston, J.F. (1992) *Financial Theory and Corporate Policy*, 3rd edn, Addison-Wesley, Reading MA.
Cosmides, L. and Tooby, J. (1992) *Cognitive Adaptions for Social Exchange*, Oxford University Press, Oxford.
Cyert, R.M. and March, J.D. (1992) *A Behavioral Theory of the Firm*, 2nd edn, Blackwell, Cambridge MA.
de Finetti, B. (1962) Does it make sense to speak of 'good probability appraisers'? In *The Scientist Speculates: An Anthology of Partly-Baked Ideas*, Good, I.J. (ed.), John Wiley & Sons, Inc., New York, pp. 357–363.
de Finetti, B. (1972) *Probability, Induction and Statistics*, John Wiley & Sons, Inc., New York.
de Finetti, B. (1974) *Theory of Probability*, John Wiley & Sons, Inc., New York.

de Groot, M.H. (1970) *Optimal Statistical Decisions*, McGraw-Hill, New York.
Dewooght, J. (1998) Model uncertainty and model inaccuracy. *Reliability Engineering and System Safety*, **59**: 171–185.
Douglas, E.J. (1983) *Managerial Economics: Theory, Practice and Problems*, 2nd edn, Prentice Hall, Englewood Cliffs NJ.
Douglas, M. and Wildavsky, A. (1982) *Risk and Culture*, University of California Press, Berkeley CA.
Draper, D. (1995) Assessment and propagation of model uncertainty. *Journal of the Royal Statistical Society*, **57**: 45–97.
Fischhoff, B., Lichtenstein, S., Slovic, P., Derby, S. and Keeney, R. (1981) *Acceptable Risk*, Cambridge University Press, Cambridge.
French, S. and Insua, D.R. (2000) *Statistical Decision Theory*, Arnold, London.
French, S., Bedford, T. and Atherton, E. (2002) Supporting ALARP decision-making by cost-benefit analysis and multi-attribute utility theory. *Journal of Risk Research*.
Geisser, S. (1993) *Predictive Inference: An Introduction*, Chapman & Hall, New York.
Good, I.J. (1950) *Probability and Weighing of Evidence*, Griffin, London.
Good, I.J. (1983) *Good Thinking: The Foundations of Probability and Its Applications*, University of Minnesota Press, Minneapolis MN.
Haimes, Y.Y. (1998) *Risk Modeling, Assessment and Management*, John Wiley & Sons, Inc., New York.
Helton, J.C. and Burmaster, D.E. (eds) (1996) *Reliability Engineering and System Safety*, special issue on treatment of aleatory and epistemic uncertainty.
Henley, E.J. and Kumamoto, H. (1981) *Reliability Engineering and Risk Assessment*, Prentice Hall, Englewood Cliffs NJ.
Hertz, D.B. and Thomas, H. (1983) *Risk Analysis and its Applications*, John Wiley & Sons, Inc., New York.
Hoffman, F.O. and Kaplan, S. (1999) Beyond the domain of direct observation: how to specify a probability distribution that represents the state of knowledge about uncertain inputs. *Risk Analysis*, **19**: 131–134.
Hood, C. and Jones, D. (eds) (1996) *Accident and Design*, UCL Press, London.
Høyland, A. and Rausand, M. (1994) *System Reliability Theory*, John Wiley & Sons, Inc., New York.
Hull, J.C. (1980) *The Evaluation of Risk in Business Investment*, Pergamon, New York.
ISO (2002) Risk management vocabulary. International Organization for Standardization ISO/IEC Guide 73.
Janis, I. and Mann, L. (1977) *Decision Making*, Free Press, New York.
Jordanger, I. (1998) Value-oriented management of project uncertainties. Paper presented at the IPMA World Congress, Ljubljana.
Kadane, J.B. (1993) Several Bayesian a review. *Test*, **2**: 1–32.
Kahneman, D., Slovic, P. and Tversky, A. (eds) (1982) *Judgement under Uncertainty: Heuristics and Biases*, Cambridge University Press, New York.
Kaplan, S. (1991) Risk assessment and risk management – basic concepts and terminology. In *Risk Management: Expanding Horizons in Nuclear Power and Other Industries*, Hemisphere, Boston MA, pp. 11–28.
Kaplan, S. (1992) Formalism for handling phenomenological uncertainties: the concepts of probability, frequency, variability, and probability of frequency. *Nuclear Technology*, **102**: 137–142.
Kaplan, S. and Burmaster, D. (1999) Foundations: how, when, why to use all of the evidence. *Risk Analysis*, **19**: 55–62.

Kaplan, S. and Garrick, B.J. (1981) On the quantitative definition of risk. *Risk Analysis*, **1**: 11–27.
Karni, E. (1996) Probabilities and beliefs. *Journal of Risk and Uncertainty*, **13**: 249–262.
Kayaloff, I.J. (1988) *Export and Project Finance*, Euromoney, Bath Press.
Keeney, R.L. (1992) *Value-Focused Thinking*, Harvard University Press, Cambridge MA.
Keeney, R.L. and Raiffa, H. (1976) *Decisions with Multiple Objectives*, Cambridge University Press, Cambridge.
Keynes, J.M. (1921) *Treatise on Probability*, Macmillan, London.
Klein, G. and Crandall, B.W. (1995) The role of mental simulation in problem solving and decision making. In *Local Applications of the Ecological Approach to Human Machine Systems*, Vol. 2, Hancock, P. et al. (eds), Erlbaum, Hillsdale NJ, pp. 324–358.
Klinke, A. and Renn, O. (2001) Precautionary principle and discursive strategies: classifying and managing risks. *Journal of Risk Research*, **4**: 159–173.
Klovning, J. and Nilsen, E.F. (1995) Quantitative environmental risk analysis. SPE conference paper 30686, Dallas TX, 22-25/10-1995.
Koller, G. (1999a) *Risk Assessment and Decision Making in Business and Industry*, CRC Press, New York.
Koller, G. (1999b) *Risk Modeling and Determining Value and Decision Making: A Practical Guide*, CRC Press, New York.
Koopman, B.O. (1940) The bases of probability. *Bulletin of the American Mathematical Society*, no. 46. Reprinted in Kyburg and Smokler (1980).
Kørte, J., Aven, T. and Rosness, R. (2002) On the use of risk analysis in different decision settings. Paper presented at ESREL 2002, Lyon.
Kristensen, V., Aven, T. and Ford, D. (2003) A safety management framework based on risk characterization. Submitted.
Kyburg, H.E. Jr and Smokler, H.E. (1980) *Studies in Subjective Probability*, Krieger, New York.
Lad, F. (1996) *Operational Subjective Statistical Methods*, John Wiley & Sons, Inc., New York.
Levy, H. (1998) *Stochastic Dominance*, Kluwer Academic, Bosten MA.
Levy, H. and Sarnat, M. (1972) Safety first – an expected utility principle. *Journal of Financial and Quantitative Analysis*, **7**: 1829–1834.
Levy, H. and Sarnat, M. (1990) *Capital Investment and Financial Decisions*, Prentice Hall, New York.
Lichtenstein, S., Slovic, P., Fischhoff, B., Layman, M. and Combs, B. (1978) Judged frequency of lethal events. *Journal of Experimental Psychology, Human Learning and Memory*, **4**: 551–578.
Lindblom, C. (1995) The science of muddling through. In *Public Policy*, Theodoulou, S.Z. (ed.), Prentice Hall, Englewood Cliffs NJ, pp. 113–127.
Lindley, D.V. (1978) The Bayesian approach. *Scandinavian Journal of Statistics*, **5**: 1–26.
Lindley, D.V. (1982) Scoring rules and the inevitability of probability. *International Statistical Review*, **50**: 1–26.
Lindley, D.V. (1985) *Making Decisions*, John Wiley & Sons, Inc., New York.
Lindley, D.V. (2000) The philosophy of statistics. *The Statistician*, **49**: 293–337.
Lindley, D.V., Tversky, A. and Brown, R.V. (1979) On the reconciliation of probability assessment. *Journal of the Royal Statistical Society A*, **142**: 146–180.
Madsen, H.O., Krenk, S. and Lind, N.C. (1986) *Methods of Structural Safety*, Prentice Hall, London.
March, J.D. and Simon, H.A. (1958) *Organizations*, John Wiley & Sons, Inc., New York.

Meeker, W.Q. and Escobar, L.A. (1998) *Statistical Methods for Reliability Data*, John Wiley & Sons, Inc., New York.
Melchers, R.E. (1987) *Structural Reliability Analysis and Prediction*, Ellis Horwood, Chichester, UK.
Mendel, M.B. (1994) Operational parameters in Bayesian models. *Test*, **3**: 195–206.
Mintzberg, H. (1973) *The Nature of Managerial Work*, Harper Collins, New York.
Moore, P. (1983) *The Business of Risk*, Cambridge University Press, Cambridge.
Morgan, M.G. and Henrion, M. (1990) *Uncertainty: A Guide to Dealing with Uncertainty in Qualitative Risk and Policy Analysis*, Cambridge University Press, Cambridge.
Mosleh, A. and Apostolakis, G. (1986) The assessment of probability distributions from expert opinions with an application to seismic fragility curves. *Risk Analysis*, **6**: 447–461.
Mosleh, A. and Bier, V.M. (1996) Uncertainty about probability: a reconciliation with the subjectivist view. *IEEE Transactions on Systems, Man and Cybernetics*, **26**: 303–310.
Mosleh, A., Siu, N., Smidts, C. and Lui, C. (1994) Model uncertainty: its characterization and quantification. Proceedings of Workshop 1, Advanced Topics in Risk and Reliability Analysis. Prepared for the Nuclear Regulatory Commission, NUREG/CP-0138.
Moyer, R., McGuigan, J. and Kretlow, W. (1995) *Contemporary Financial Management*, 6th edn, West Publishing, New York.
Murphy, A.H. and Winkler, R.L. (1992) Diagnostic verification of probability forecast. *International Journal of Forecasting*, **7**: 435–455.
Myers, S.C. (1976) Postscript: using simulation for risk analysis, In *Modern Developments in Financial Management*, S.C. Myers (ed.), Praeger, New York.
Natvig, B. (1997) How did Thomas Bayes think? (In Norwegian) *Utposten. Blad for allmenn- og samfunnsmedisin*, **26**: 348–354.
Nevitt, P.K. (1989) *Project Financing*, 5th edn, Euromoney, Bath Press.
Nilsen, T. and Aven, T. (2003) Models and model uncertainty in the context of risk analysis. *Reliability Engineering and System Safety*, **79**: 309–317.
Nilsen, T., Aven, T. and Jakobsen, G.S. (2000) Requirements and principles for cause analysis in QRA – with application. In *Proceedings of the European Safety and Reliability Conference*, pp. 641–646.
NORSOK (1999) Regularity management and reliability technology. Z-016, Norwegian Technology Standards Institution, Oslo.
NORSOK (2001) Risk and emergency preparedness analysis. Z-013, Norwegian Technology Standards Institution, Oslo.
Okrent, D. and Pidgeon, N. (eds) (1998) *Reliability Engineering and System Safety*, special issue on risk perception versus risk analysis.
Otway, H. and Winterfeldt, D. (1992) Expert judgement in risk analysis and management: process, context, and pitfalls. *Risk Analysis*, **12**: 83–93.
Pape, R.P. (1997) Developments in the tolerability of risk and the application of ALARP. *Nuclear Energy*, **36**: 457–463.
Parry, G.W. and Winter, P.W. (1981) Characterization and evaluation of uncertainty in probabilistic risk analysis. *Nuclear Safety*, **22**: 28–42.
Perrow, C. (1984) *Normal Accidents*, Basic Books, New York.
Pidgeon, N.F. and Beattie, J. (1998) The psychology of risk and uncertainty. In *Handbook of Environmental Risk Assessment and Management*, Calow, P. (ed.), Blackwell Science, London, pp. 289–318.
Popper, K. (1959) *The Logic of Scientific Discovery*, Hutchinson, London.

Raiffa, H. (1968) *Decision Analysis*, Addison-Wesley.
Ramsberg, J.A. and Sjöberg, L. (1997) The cost-effectiveness of lifesaving interventions in Sweden. *Risk Analysis*, **17**: 467–478.
Ramsey, F. (1926) *Truth and Probability*. Reprinted in Kyburg and Smokler (1980).
Rasmussen, J. (1986) *Information Processing and Human-Machine Interaction*, North-Holland, Amsterdam.
Rasmussen, J. (1991) Event analysis and the problem of causality. In *Distributed Decision Making*, Rasmussen, J., Brehmer, B. and Leplat, J. (eds) John Wiley & Sons, Inc., New York, pp. 247–256.
Rasmussen, J. (1997) Risk management in a dynamic society: a modelling problem. *Safety Science*, **27**: 183–213.
Reason, J. (1990) *Human Error*, Cambridge University Press, Cambridge.
Reason, J. (1997) *Managing the Risks of Organizational Accidents*, Ashgate, Aldershot, UK.
Rosness, R. and Hovden, J. (2001) From power games to hot cognition – a contingency model of safety related decision-making. Paper presented at the Workshop on Decision Making under Uncertainty, Molde, 19–21 May 2001.
Ross, S. (1993) *Introduction to Probability Models*, 5th edn, Academic Press, New York.
Saaty, T.L. and Vargas, L.G. (2001) *Models, Methods, Concepts and Application of the Analytical Hierarchy Process*, Kluwer Academic, London.
Samurcay, R. and Rogalski, J. (1991) A method for tactical reasoning in emergency management. In *Distributed Decision Making*, Brehmer, B. Rasmussen, J. and Leplat, J., (eds) John Wiley & Sons, Inc., New York, pp. 287–297.
Sandøy, M. and Aven, T. (2003) Application of sensitivity analysis for a risk analysis tool for blowouts. ESREL 2003, Maastricht, 15–18 June.
Savage, L.J. (1962) Subjective probability and statistical practice. In *The Foundations of Statistical Inference*, John Wiley & Sons, Inc., New York.
Schulman, P.R. (1995) Nonincremental policy making. In *Public Policy*, Theodoulou, S.Z. (ed.), Prentice Hall, Englewood Cliffs NJ, pp. 129–137.
Shrader-Frechette, K.S. (1991) *Risk and Rationality*, University of California Press, Berkeley CA.
Simon, H.A. (1957a) *Models of Man*, John Wiley & Sons, Inc., New York.
Simon, H.A. (1957b) *Administrative Behavior: A Study of Decision-Making Processes in Administrative Organization*, 2nd edn, Macmillan, New York.
Singpurwalla, N.D. (1988) Foundational issues in reliability and risk analysis. *SIAM Review*, **30**: 264–282.
Singpurwalla, N.D. (2000) Warranty contracts and equilibrium probabilities. In *Statistical Science in the Courtroom*, Gastwirth, J.L. (ed.), Springer-Verlag, pp. 363–377.
Singpurwalla, N.D. (2002) Some cracks in the empire of chance. *International Statistical Review*, **70**: 53–79.
Singpurwalla, N.D. and Wilson, S.P. (1999) *Statistical Methods in Software Engineering*, Springer-Verlag, New York.
Slovic, P. (1998) The risk game. *Reliability Engineering and System Safety*, **59**: 73–77.
Spizzichino, F. (2001) *Subjective Probability Models for Lifetimes*, Chapman & Hall, New York.
Stern, P.C. and Fineberg, H.V. (eds) (1996) *Understanding Risk*, National Academy Press, Washington DC.
Tarantola, S. and Saltelli, A. (eds) (2003) *Reliability Engineering and System Safety*, special issue on sensitivity analysis, Vol. 97, no 2.

Tenga, T.O., Adams, M.E., Pliskin, J.S., Safran, D.G., Siegel, J.E., Weinstein, M.C. and Graham, J.D. (1995) Five hundred life-saving interventions and their cost-effectiveness, *Risk Analysis*, **15**: 369–390.
Toft-Christensen, P. and Baker, M.J. (1982) *Structural Reliability Theory and Its Applications*, Springer-Verlag, New York.
Tversky, A. and Kahneman, D. (1974) Judgments under uncertainty: heuristics and biases. *Science*, **185**: 1124–1131.
UKOOA (1999) A framework for risk related decision support. Industry guidelines, UK Offshore Operators Association, May 1999.
Vatn, J. (1998) A discussion of the acceptable risk problem. *Reliability Engineering and System Safety*, **61**: 11–19.
Vinnem, J.E. (1999) *Offshore Risk Assessment*, Kluwer Academic, London.
Vinnem, J.E., Tveit, O., Aven, T. and Ravnås, E. (2002) Use of risk indicators to monitor trends in major hazard risk on a national level. In *Proceedings of the European Safety and Reliability Conference*, pp. 512–518.
von Neumann, J. and Morgenstern, O. (1944) *Theory of Games and Economics*, Princeton University Press, Princeton NJ.
von Winterfeldt, D. and Edwards, W. (1986) *Decision Analysis and Behavioral Research*, Cambridge University Press, Cambridge.
Vose, D. (2000) *Risk Analysis*, 2nd edn, John Wiley & Sons, Inc., New York.
Watson, S.R. (1994) The meaning of probability in probabilistic safety analysis. *Reliability Engineering and System Safety*, **45**: 261–269.
Watson, S.R. and Buede, D.M. (1987) *Decision Synthesis*, Cambridge University Press, New York.
Webster's Dictionary (1989) *Encyclopedic Unabridged Dictionary of the English Language*, Gramercy Books, New York.
Winkler, R.L. (1968) Good probability assessors. *Journal of Applied Meteorology*, **7**: 751–758.
Winkler, R.L. (1986) On good probability appraisers. In *Bayesian Inference and Decision Techniques*, Goel, P. and Zellner, A. (eds), Elsevier London, pp. 265–278.
Winkler, R.L. (1996a) Uncertainty in probabilistic risk assessment. *Reliability Engineering and System Safety*, **54**: 127–132.
Winkler, R.L. (1996b) Scoring rules and the evaluation of probabilities. *Test*, **5**: 1–60.
Wright, G. and Ayton, P. (1994) *Subjective Probability*, John Wiley & Sons, Inc., New York.
Zio, E. and Apostolakis, G. (1996) Two methods for the structured assessment of model uncertainty by experts in performance assessment of radioactive waste repositories. *Reliability Engineering and System Safety*, **54**: 225–241.

Index

Acceptable risk problem, 113
Accident statistics, 7
Accountability, 75
Actuarial risk, 15
AHP, 127
ALARP, 22, 39, 107, 138
Alternating renewal process, 57
Ambiguity, 41, 51
Analytical hierarchy process, 127
Authority level, 133

Background information, 50, 81, 87, 89, 93, 146, 150, 165
Bayes' factor, 173
Bayes' theorem, 153
Bayesian analysis, 86, 89, 146
Bayesian approach, x, 37, 42, 62, 72, 75, 91, 164
Bayesian decision analysis, xii, 101, 173
Bayesian inference, 171
Bayesian statistics, 80
Bayesian updating, 38, 72, 76, 92
Behavioural decision research, 41
Bernoulli trial, 159
Best estimate, 12, 26
Beta distribution, 88, 163
Beta-binomial distribution, 88, 163
Binomial distribution, 159, 169
Birnbaum's measure, 89
Blunt end, 132
Bounded rationality, 105, 135, 142

Calibration, 65
CAPM, capital asset pricing model, 31
Central limit theorem, 163
Chance, 52, 79

Characteristic lifetime, 161
Chi-square distribution, 162
Classical approach, 36
Coherence, 64
Common cause, 87
Conditional expectation, 157
Conditional probability, 153, 157
Confidence interpretation, 82, 88
Confidence interval, 16, 31, 168
Confidence measure, 82
Conjugate distributions, 83, 172
Consensus, 68, 75, 103, 106, 124, 136
Consequence, 175
Convolution, 55, 158
Correlation coefficient, 31, 59, 84, 157
Cost risk analysis, 30
Cost-benefit analysis, xii, 39, 99, 107, 109, 119, 136, 146
Counting process, 165
Covariance, 157
Credibility interval, 172
Crises and emergency management, 139

de Finetti's representation theorem, 80
Decision aid, 137
Decision analysis, xii, 98, 127, 138, 173
Decision node, 118
Decision setting, 132
Decision tree, 118
Decision-making, 4, 30, 39
    organizations, 142
Decision-making model, 97
Degree of belief, 41, 150
Delay effect, 128
Deliberation, 106, 143
Dependency modelling, 84, 86, 87
Descriptive approach, 95
Discount rate, 32, 34, 59

Foundations of Risk Analysis Branka Vucetic and Jinhong Yuan
© 2003 John Wiley & Sons, Ltd ISBN: 0-471-49548-4

Distribution
  beta, 163
  beta-binomial, 163
  binomial, 159
  chi-square, 162
  exponential, 160
  gamma, 162
  geometric, 159
  lognormal, 164
  multivariate normal, 164
  normal, 163
  Poisson, 160
  triangular, 163
  uniform, 160
  Weibull, 161
Diversification, 32, 59

Empirical control, 75, 90
Empirical distribution function, 166
Environmental organizations, 110
Estimation, 54
  non-parametric, 166
Event, 175
Event tree, 12
Example
  accident risk, 60, 69, 79, 91, 106
  business and project management, 57
  cost risk analysis, 52, 69, 83
  criminal law, 78
  health risk, 75, 91, 116, 142
  medical treatment, 4
  multi-attribute utility, 124
  offshore development project, 2, 96, 120
  offshore safety risk analysis, 11
  production risk, 55, 69, 85, 113
  reliability target, 114
  removal of plant, 108
  risk assessment, national sector, 122
  stock market, 3
  warranties, 119, 143
Exchangeability, 80, 156
Expectation, 155
  conditional, 157
Expected utility, 101, 110
Exponential distribution, 25, 86, 160, 167, 172
Exponential transform, 126

Failure rate, 26, 57, 161
Fairness, 75

FAR, 12, 17, 61, 106
Fault tree, 25
Fictional parameter, 38, 54, 62, 80, 91
Formal expert elicitation, 74, 146
Fuzzy logic, xii

Gamma distribution, 162, 172
Gamma function, 162
Geometric distribution, 159
Goodness of fit tests, 167
Group decision-making, 103, 106, 142

Hazard, 123
Hazard (cumulative failure rate), 161
Hazard level, 132
Health risk, 142
Heuristics, 66, 145
Hypothesis testing, 9, 78

Ignorance, 130
Importance analysis, 20, 89
Improvement potential, 89
Independence, 57, 86, 87, 153, 156
Instrumental, xiv
Intensity process, 166
Interested party, 175

Job safety analysis, 135

Knowledge-based behaviour, 134
Kolmogorov's axioms, 152

Law of large numbers, 158
Law of total probability, 153
Least square regression, 54
Lifetime distribution, 161
Likelihood function, 76, 167
Likelihood principle, 173
Limit state function, 27
Lognormal distribution, 53, 164

Management decisions, 135
Managerial review and judgement, 98
Maximum entropy, 83
Maximum likelihood estimation, 167
Mean value, 155
Measurement errors, 64
Minimax, 173
Mitigation, 175
Model, xi, 146
Model uncertainty, 51, 71, 89

Modelling, 60, 68, 146
  offshore safety risk analysis, 13
Monotone system, 24, 86
Monte Carlo simulation, 18, 31, 55, 57, 84
Muddling through paradigm, 137
Multi-attribute analysis, 98, 105, 119, 146
Multi-attribute utility analysis, 124
Multinominal distribution, 93
Multivariate normal distribution, 27, 85, 89, 164

Near miss, 10
Nelson–Aalen estimator, 166
Neutrality, 75
Non-parametric estimation, 166
Non-systematic risk, 32
Normal distribution, 11, 29, 30, 53, 55, 58, 85, 163, 167
Normative approach, 95, 105
Notational risk, 15
NPV, 34, 100, 114, 121, 137

Observable parameter value, 69, 88
Observable quantity, ix, xi, 48, 51, 93, 145, 176
  relative frequency, 51
Odds, 47, 78
Operations, 134
Opportunity, 29

Parameter
  fictional, 38, 54, 62, 80, 91
Parametric distribution class, 54
Persistency, 128
Personal probability, xii
PLL, 12, 17
Poisson distribution, ix, 8, 12, 81, 93, 123, 160, 167, 171
Poisson process, 16, 86, 165, 168
Political decisions, 136
Portfolio theory, 31
Possibility theory, xii
Posterior distribution, 76, 117, 171
Potential of mobilization, 128
Pragmatic criterion, 64, 67
Predictability, 5
Prediction, xi, 48, 53, 56, 58, 145
Prediction interval, 31, 53, 123

Predictive Bayesian approach, xiii, 62
Predictive distribution, 19, 88, 93, 171
Preference ordering, 30
Preferential independence, 126
Prequential prediction, 173
Prescriptive approach, 95
Prior distribution, xii, 76, 79, 82, 165, 171
  improper, 83
  non-informative, 83
Probabilistic safety analysis, PSA, 6
Probability, 176
  classical interpretation, 149
  conditional, 153, 157
  personal, xii, 38
  relative frequency interpretation, 149
  subjective, xii, 38, 149
Probability assignments, 71
Probability axioms, 64
Probability model, 79, 165
Probability of frequency framework, 20, 37
Probability specification, 63
Probability verification, 64, 75
Probability wheel, 66
Propensity, 62
Pure risk, 29

Quantitative risk analysis, QRA, 6

Random nodes, 118
Random process, 165
Random quantity, 155
Random variable, 155
Randomness, 8
Rare events, 66
Rate of return, 34
Rational consensus, 75
Rationality, 30, 39, 105, 142
Real risk, 112
Refinement, 65
Regression analysis, 33, 84, 170
Reliability analysis, 24, 61, 86
Reliability block diagram, 25
Reliability model, 56
Reproducibility, 75
Residual risk, 176
Resilience, 127
Reversibility, 128
Risk, 4, 50, 176

Risk acceptance, 22, 42, 176
Risk acceptance criterion, 22, 107, 110, 176
Risk analysis, 11, 176
Risk analysis approach
   Bayesian, xiii, 38
   classical, 36
      best estimates, xiii, 12
      uncertainty analysis, xiii, 16, 89
   predictive, xiii
   predictive, Bayesian, xiii, 62
   predictive, epistemic, 62
   probability of frequency framework, 20
Risk assessment, 176
Risk aversion, 30, 126
Risk avoidance, 176
Risk communication, 107, 112, 176
Risk control, 176
Risk criteria, 176
Risk evaluation, 61, 177
Risk financing, 177
Risk identification, 177
Risk indicator, 123
Risk management, 2, 96, 131, 177
Risk management system, 177
Risk measures, 50
Risk optimization, 138, 177
Risk perception, 108, 112, 142, 177
Risk perception research, 41
Risk problem classification, 127
Risk quantification, 177
Risk reduction, 177
Risk retention, 177
Risk tolerability, 22, 42, 107
Risk transfer, 96, 178
Risk treatment, 96, 127, 178
Routine operations, 134
Rule-based behaviour, 134

Safe Job Analysis, 135
Safety function, 13
Satisficing behaviour, 105, 135
Scatter plot, 84
Science, 92
Scoring rule, 65, 90
Semantic criterion, 65
Sensitivity analysis, 20, 89, 107
Sharp end, 132
Sharpness, 65

Skill-based behaviour, 134
Social risk problem, 106
Social science, xiv, 41
Source, 178
Source identification, 178
Speculative risk, 29
Stakeholder, 127, 136, 178
Standard deviation, 156
Standardization, 68, 136, 146
Statistical decision analysis, 173
Statistical inference, 166
Statistical life, 39, 104, 126, 142
Stochastic process, 165
Structural reliability analysis, 27, 89
Structure function, 24
Subjective probability, xii
Supervisory body, 111
Syntactic criterion, 64
System reliability, 26
Systematic risk, 32

Testing hypotheses, 169
Trade-offs, 5, 99, 105, 126, 131, 138, 147
Trend analysis, 9, 122
Triangular distribution, 55, 163

Ubiquity, 128
Uncertainty, 4, 178
   aleatory, 17, 28, 37, 79, 82, 165, 171, 175
   epistemic, xi, 17, 28, 48, 79, 82, 145, 165, 171, 175
   unknown, 130
Uncertainty assessment, 63, 71
Uniform distribution, 160
Utility-based analysis, 146
Utility function, 117, 124
Utility theory, 30, 39

Vagueness, 41, 91
Value function, 126
Value of a statistical life, 104, 107, 126, 142
Variance, 156
Venn diagram, 151
Verification, 64, 75

Weibull distribution, 86, 161
Willingness to accept, 105
Willingness to pay, 105